Introduction to Light Trapping in Solar Cell and Photo-detector Devices

Introduction to Light Trapping in Solar Cell and Photo-detector Devices

Stephen J. Fonash

ELSEVIER

AMSTERDAM • BOSTON • HEIDELBERG • LONDON
NEW YORK • OXFORD • PARIS • SAN DIEGO
SAN FRANCISCO • SINGAPORE • SYDNEY • TOKYO

Academic Press is an imprint of Elsevier

Academic Press is an imprint of Elsevier
32 Jamestown Road, London NW1 7BY, UK
525 B Street, Suite 1800, San Diego, CA 92101-4495, USA
225 Wyman Street, Waltham, MA 02451, USA
The Boulevard, Langford Lane, Kidlington, Oxford OX5 1GB, UK

First edition 2015

British Library Cataloguing in Publication Data
A catalogue record for this book is available from the British Library

Library of Congress Cataloging-in-Publication Data
A catalog record for this book is available from the Library of Congress

ISBN: 978-0-12-416649-3

For information on all Academic Press publications
visit our website at http://store.elsevier.com/

This book has been manufactured using Print On Demand technology. Each copy is produced to
order and is limited to black ink. The online version of this book will show color figures where
appropriate.

Working together
to grow libraries in
developing countries

www.elsevier.com • www.bookaid.org

DEDICATION

To Evan, Nina, Leo, and Alec – and their parents – and
to their grandmother Joyce

TABLE OF CONTENTS

PREFACE

I have been very involved in light trapping and carrier collection in solar cells for quite some time. While these two phenomena are really very intertwined, as I discussed in my 2010 Elsevier book "Solar Cell Device Physics," this work concentrates on the light-trapping aspect. The attempt here is to look systematically at light trapping and to explore, in a short and concise manner, how it can be accomplished. Polarization effects are only very briefly mentioned since the intent is to capture the essence of light trapping. My interest is solar cells but photo-detectors are mentioned explicitly here and there and the applicability of the discussion to such devices should be obvious.

I am very indebted to Drs Atilla Ozgur Cakmak and Nghia Dai Nguyen here at Penn State for their support and aid in this project. Both contributed to the computer modeling results used, and Dr Cakmak was particularly helpful in looking at special computer cases and in proofreading the text. My appreciation also goes to Renee Lindenberg for her assistance and patience in getting this book together.

The level intended for this work is that of engineering and science seniors, practicing engineers, and first-year graduate students. I tried to go easy on the mathematics and to concentrate on bare-bones physics. Hopefully I did that effectively.

CHAPTER *1*

A Brief Overview of Phenomena Involved in Light Trapping

Light trapping is the capturing of as many photons as possible from an impinging electro-magnetic (E-M) wave with the objective of generating heat or charge carriers, excitons, or both [1]. For our purposes, the "light" being trapped may lie anywhere in that part of the E-M spectrum extending from the infrared to the ultraviolet. This range encompasses interesting frequencies, such as the photon-rich part of the solar spectrum, the emission and absorption frequencies of living things, and astronomically useful frequencies. Trapping of this light is vital to many applications, including sensing [2,3], photovoltaics [1], photo-electrochemistry [1,4], solar fuel production [5], and thermal photovoltaics [6]. We limit ourselves in this text to light trapping for production of charge carriers, excitons, or both. Some discussions break light trapping into light capture and light trapping. With the thin devices and, in particular, the thin photon absorbers now used in high-speed detectors and advanced solar cells today, it is rather artificial to separate light capture and light trapping. In this text, they are treated as one light trapping process, as seen in Figure 1.1.

When light enters a structure, we will find that it can reside in a number of different modes: radiation modes, trapped traveling modes (guided modes[1] and Bloch modes[2]), and trapped localized modes (Mie modes[3] and plasmonic modes) as well as hybridizations[4] of these. We will develop

---

[1] Waveguide modes are waves that propagate in the open direction(s) of a waveguide. They are bound in the waveguide in the direction (1-D) or in the directions (2-D) perpendicular to the open direction(s).

[2] If a structure exhibits periodicity in the open direction(s) of a waveguide-like structure, the waves in the device must be Bloch modes. They too move in the open direction(s) and are bound in the waveguide.

[3] A Mie mode is one in which the electric field is localized around or in a wavelength-scale feature.

[4] Hybridization is used here as it is in quantum mechanics; it is a mixture of states with the same energy. For example, in quantum mechanics carbon, in its diamond configuration, has bonding which is the result of the specific hybridization (specific mixing) of its 2s and 2p wave functions.

Introduction to Light Trapping in Solar Cell and Photo-detector Devices. DOI: 10.1016/B978-0-12-416649-3.00001-4

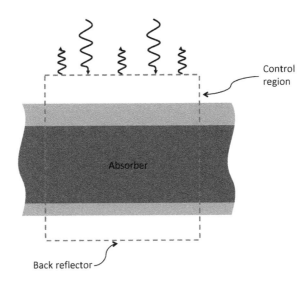

Fig. 1.1. A control region showing the definition of light trapping used in this text. Light trapping endeavors to minimize the light energy propagating away at the top control surface. Normal impingement onto a general structure is assumed in this figure.

familiarity with all of this terminology in Chapter 2. For now, we can note that light trapping may be thought of as the task of populating some or all of these various modes possible in a structure with photons.

The optical phenomena that provide the key tools for light trapping are listed in Table 1.1 along with their definitions, as utilized in this monograph. After reviewing these tools in the remainder of this chapter, we will move in Chapter 2 to examining those modes in which photons may be stored in a structure. In Chapter 3, we will look at various light-trapping structures. As may be seen from Figure 1.2a–d, these structures can range from devices with simple ¼ wavelength antireflection coatings (ARCs) to devices with honeycomb gratings. When needed, we will consider polarization effects. Chapter 4 provides final comments on light-trapping structures and approaches.

In our brief discussion of the phenomena of Table 1.1, we occasionally will use the approximation of geometrical optics.[5] However, our approach will generally be to employ physical optics and, when needed, to implement it with full numerical solutions of Maxwell's equations.

---

[5] The correct formulation of optics is physical (wave) optics. Geometrical (ray tracing) optics is an approximation, which is often useful when the wavelength is small compared to the characteristic lengths involved in a structure.

Table 1.1 Light-Trapping Phenomena and Definitions	
Interference	The phenomenon whereby two or more E-M waves existing at a point constructively or destructively add together to some degree at that point.
Scattering	The result of impinging E-M waves bouncing off of objects by being absorbed and emitted. Material property dependent. In general, can be elastic or inelastic.[6]
Reflection	The result of some portion of an impinging E-M wave being scattered backwards. Material property dependent. Generally taken as elastic.[6]
Diffraction	The result of impinging E-M waves bouncing off of objects by being absorbed, effectively instantaneously emitted, and constructively interfering in certain specific directions. Taken as elastic.[6]
Plasmonics	The result of an impinging E-M wave being absorbed by the extremely numerous electrons of a metal thereby exciting an oscillating plasma. This plasma dissipates energy through electron collisions and also reradiates an E-M scattered wave. Material property dependent. Over all, inelastic.[6]
Refraction	The result of an impinging E-M wave changing direction and wavelength due to a change in the transmission medium through which it is passing. Material property dependent. Taken as elastic.[6]

1.1 INTERFERENCE

Interference is listed first in Table 1.1 because it is the defining trait of waves. This basic behavior can be considered by watching two electric field waves coming from two plane wave sources (source 1 and source 2). If we examine these waves at some point which is $\vec{r_1}$ from source 1 of the first wave and $\vec{r_2}$ from source 2 of the second wave, then wave 1 at this point is of the form $\left[A_x\hat{i} + A_y\hat{j} + A_z\hat{k} \right]e^{i[\vec{k_1}\vec{r_1}-\omega_1 t]}$ with k-vector $\vec{k_1}$ and angular frequency ω_1, whereas, wave 2 is of the form $\left[B_x\hat{i} + B_y\hat{j} + B_z\hat{k} \right]e^{i[\vec{k_2}\vec{r_2}-\omega_2 t]}$ with k-vector $\vec{k_2}$ and angular frequency ω_2. As we know, these waves simply add up at our point of interest to

$$\vec{\xi} = \left\{ A_x e^{i[\vec{k_1}\vec{r_1}-\omega_1 t]} + B_x e^{i[\vec{k_2}\vec{r_2}-\omega_2 t]} \right\}\hat{i}$$
$$+ \left\{ A_y e^{i[\vec{k_1}\vec{r_1}-\omega_1 t]} + B_y e^{i[\vec{k_2}\vec{r_2}-\omega_2 t]} \right\}\hat{j} \qquad (1.1)$$
$$+ \left\{ A_z e^{i[\vec{k_1}\vec{r_1}-\omega_1 t]} + B_z e^{i[\vec{k_2}\vec{r_2}-\omega_2 t]} \right\}\hat{k}$$

where $\vec{\xi}$ is the total electric field. The total E-M energy density U present at our arbitrary point is proportional to the square of the magnitude of the electric field [8,9]; i.e.,

---

[6] Elastic in this case means there is no net conversion of E-M energy into other forms; inelastic is the converse.

$$U = \varepsilon \vec{\xi} \cdot \vec{\xi}^* \tag{1.2}$$

where $\vec{\xi}^*$ is the complex conjugate of $\vec{\xi}$ and ε the permittivity at the point in question. The total photon density is therefore proportional to this U. Using Eq. 1.2 it follows that,

$$U = \varepsilon \left[\begin{array}{l} \left\{ \begin{array}{l} A_x^2 + B_x^2 + A_x B_x e^{-i[\vec{k_1}\vec{r_1}-\omega_1 t]} e^{i[\vec{k_2}\vec{r_2}-\omega_2 t]} \\ + A_x B_x e^{i[\vec{k_1}\vec{r_1}-\omega_1 t]} e^{-i[\vec{k_2}\vec{r_2}-\omega_2 t]} \end{array} \right\} \\ + \left\{ \begin{array}{l} A_y^2 + B_y^2 + A_y B_y e^{-i[\vec{k_1}\vec{r_1}-\omega_1 t]} e^{i[\vec{k_2}\vec{r_2}-\omega_2 t]} \\ + A_y B_y e^{i[\vec{k_1}\vec{r_1}-\omega_1 t]} e^{-i[\vec{k_2}\vec{r_2}-\omega_2 t]} \end{array} \right\} \\ + \left\{ \begin{array}{l} A_z^2 + B_z^2 + A_z B_z e^{-i[\vec{k_1}\vec{r_1}-\omega_1 t]} e^{i[\vec{k_2}\vec{r_2}-\omega_2 t]} \\ + A_z B_z e^{i[\vec{k_1}\vec{r_1}-\omega_1 t]} e^{-i[\vec{k_2}\vec{r_2}-\omega_2 t]} \end{array} \right\} \end{array} \right] \tag{1.3a}$$

This expression can be somewhat simplified to

$$U = \varepsilon \left[\begin{array}{l} \left\{ A_x^2 + B_x^2 + A_y^2 + B_y^2 + A_z^2 + B_z^2 \right\} \\ + \left\{ 2A_x B_x + 2A_y B_y + 2A_z B_z \right\} \cos\left(\vec{k_1} \cdot \vec{r_1} - \vec{k_2} \cdot \vec{r_2} + \omega_2 t - \omega_1 t \right) \end{array} \right] \tag{1.3b}$$

Equation 1.3b is interesting because it shows that the quantity $\bar{U}$, energy density averaged over time and space, has the value that we would probably expect; i.e., it is

$$\bar{U} = \varepsilon \left[\left\{ A_x^2 + B_x^2 + A_y^2 + B_y^2 + A_z^2 + B_z^2 \right\} \right] \tag{1.4}$$

The cosine term in Eq. 1.3b also shows that interference between these waves can increase or decrease the expected average energy density $\bar{U}$ at different places and times by as much as $2A_x B_x + 2A_y B_y + 2A_z B_z$. For example, if the two waves have the same frequency ω and the same amplitudes A for all components, and if the phase difference $\vec{k_1} \cdot \vec{r_1} - \vec{k_2} \cdot \vec{r_2}$ works out to 0, 2π, etc. at our observation point, then there is total constructive interference and $U = 12A^2$. If the phase difference $\vec{k_1} \cdot \vec{r_1} - \vec{k_2} \cdot \vec{r_2}$ of these two waves works out to π, 3π, etc. at another observation point, then U is zero at that point because total destructive interference has occurred. Of course, everything in between these two extremes can occur.

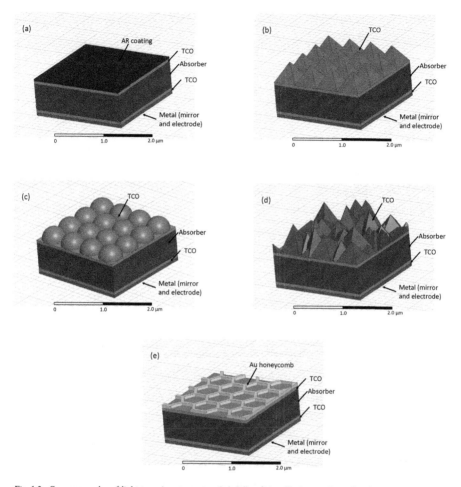

Fig. 1.2. Some examples of light-trapping structures: (a) A "traditional" planar solar cell with an ARC; (b) a solar cell with a periodically arranged pyramidal TCO array; (c) a solar cell with a periodic array of TCO domes; (d) a solar cell with a randomly textured TCO layer; and (e) a solar cell with a honeycomb nanoscale grating structure.

1.2 SCATTERING

Scattering is the general name for light "bouncing" off of some object in its path. Some examples of scattering are seen in Figure 1.3. The actual scattering event can consist of absorption, emission, and propagation away from the object. In general, the process may be elastic or inelastic, although, except for plasmonics, we will generally limit our discussion to elastic processes. For most materials, the emission following absorption

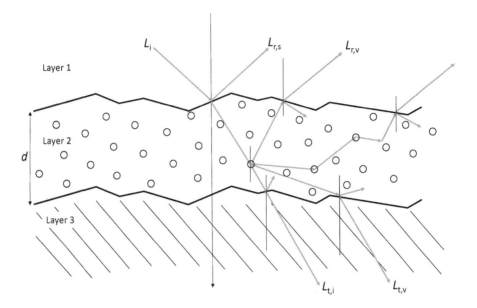

Fig. 1.3. A schematic representation of various types of scattering, including scattering off of nanoparticles in layer 2, taking place in a three-layered structure with textured interfaces. A geometrical optics ray picture is used for convenience.

involves an electron relaxation process in the scatterer, which takes place in less than 10^{-8} s [8]. Usually, this emission time is so short that its granularity [2] is not considered in light trapping

1.3 REFLECTION

Reflection is the scattering of a wavefront at the boundary between two different media so that the wavefront returns into the medium from which it originated with its photon energy E, magnitude of the total momentum, and momentum parallel to the boundary conserved. Equivalently, its wave angular frequency ω, wavelength, and wave vector component parallel to the boundary are conserved [8]. These conditions imply that the angle of incidence must be equal to the angle of reflection.

1.3.1 Reflection at Interfaces
1.3.1.1 Specular and Diffuse Reflection
When reflection occurs at an essentially planar interface with roughness that is small in height and extent compared to the wavelength of the impinging light, reflected light from across this interface is essentially in

phase and the angle of incidence θ_i must equal the angle of reflection θ_f. This situation is termed specular reflection [9]. On the other hand, when interface roughness is of the order of the wavelength in size, an assemblage of such features causes an angular distribution of the reflection coming off and this distribution can be over a broad range of angles. When the distribution is over a broad range of angles, the situation is called diffuse reflection [8,9]. An air-medium interface (surface) with perfect diffuse reflectance is defined to be one with the same apparent brightness regardless of the observer's angle of view of the surface. Such perfect diffuse reflectance is termed Lambertian reflectance [10].

1.3.2 Reflection Within Structures
1.3.2.1 Perfectly Diffuse Reflection
In this section, we ask the following question: how much light trapping can we obtain in a structure due to the internal "bouncing around" possible with diffuse light reflection. Obviously, such "bouncing around" is an approach to light trapping. Figure 1.4 is helpful in answering this. It depicts a layer with textured (i.e., randomly rough) top and bottom interfaces, each of which is assumed to give perfectly diffuse reflection. This combined diffuse reflection totally randomizes the internal light

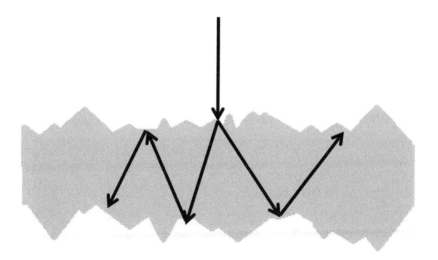

Fig. 1.4. A layer with randomly rough (textured) top and bottom interfaces. It is assumed that these interfaces both cause perfectly diffuse reflection. A ray-tracing picture shows the light direction being randomized inside the layer.

paths. Due to the "bouncing around," light trapping builds up inside the structure until it can produce, in steady state, light loss out from the top of the structure which equals light impingement into the top minus steady-state absorption in the structure. Complete bouncing around and mixing up of the light can be looked upon as populating all the radiation and trapped modes in the device [7]. Pursuing a detailed balance analysis shows that having light trapping in Figure 1.4 due to a ray optics "bouncing around" picture leads ideally to a $4n^2$ enhancement in absorptance (defined as absorption/incoming intensity) for this structure [11]. The $4n^2$ factor, frequently called the Yablonovitch limit, is derived assuming perfectly diffuse reflection at both interfaces in Figure 1.4, as well as weak absorption in the absorber. This derivation is presented in Appendix A. While the absorptance can never exceed unity, later analyses have shown that for some structures and wavelengths, the enhancement factor can be larger than $4n^2$ [12]. However, the $4n^2$ factor is a straightforward, useful upper limit for absorption enhancement against which other light-trapping approaches may be measured. An in-depth discussion of this $4n^2$ limit, which we will utilize in Chapter 3, is given in Appendix A.

1.3.2.2 Fabry–Perot Reflection

We now embark upon discussing a reflection phenomenon that can occur due to specular reflection of radiation modes at interfaces of, and the resulting interfaces within, a structure such as that seen in Figure 1.5. This phenomenon bears the designation Fabry–Perot reflection [13]. It must be carefully controlled to prevent light's escaping out of a structure and therefore loss of light-trapping possibilities. We will briefly explore

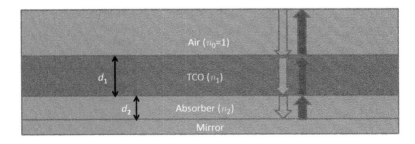

Fig. 1.5. A schematic of reflection possibilities in three interface/three layer planar structure.

which layers and wavelengths are primarily involved in Fabry–Perot reflection for the simple planar structure of Figure 1.5.

We undertake this exploration by using numerical solutions of Maxwell's equations [14]. The case being examined is that of an impinging wave polarized[7] such that the electric field $\vec{\xi}$ is normal to the layers of the structure. The resulting reflection behavior is shown in Figure 1.6 for a structure in Figure 1.5 composed of a transparent conducting oxide (TCO) layer 1 of ZnO, a layer 2 of the strong absorber PbS [15], and a back (metal) mirror reflector. These layer thicknesses are varied and their influence compared in Figure 1.6. As may be noted that for the thicknesses used, doubling the absorber thickness has little or no effect on radiation-wave reflection spectrum in what we term the first region in Figure 1.6 for the material and thicknesses utilized. On the other hand, the thinnest ZnO layer utilized suppresses the amount of ringing in the first region. This brief discussion supports the suggestion that nonabsorbing, transparent top-layer ARC materials can be useful in light trapping. In what we term the second region, both absorber and TCO thicknesses become important. The numerical results suggest that the TCO and absorber layer have a weighted joint responsibility for the Fabry–Perot peaks at longer wavelengths. In succeeding chapters, we will look into how these observations hold up for the complex structures we will encounter.

1.4 DIFFRACTION

As noted in Table 1.1, diffraction is a scattering phenomenon involving interference. It can occur at a single object or it can be a "cooperative" phenomenon involving constructive and destructive interferences arising from a number of periodically spaced objects [8]. We will use Figure 1.7 in this brief review of the latter situation since periodic multifeature diffraction will prove to be of considerable interest for light trapping.

---

[7] Polarization describes the orientation of the electric field and therefore of the magnetic field for a given E-M wave. Using the plane defined by an incoming wave and its reflected wave vector, s-polarization (also known as transverse electric (TE) polarization) has the electric field normal to this plane while p-polarization (also called transverse magnetic (TM) polarization) has the electric field lying in this plane.

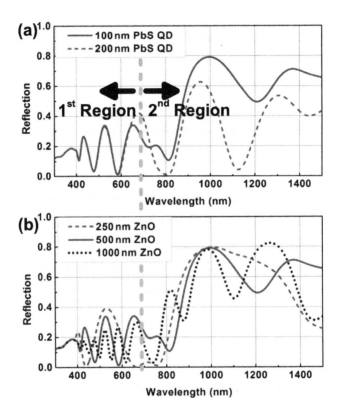

Fig. 1.6. The reflectance (reflection/incoming intensity) spectrum for a 100 nm PbS layer as the absorber and a 500 nm ZnO layer as the TCO of Fig. 1.5. Part (a) shows the effect of doubling this absorber thickness. Part (b) shows the effect of doubling this TCO thickness. The dashed curve in (b) is for halving the TCO thickness. Absorption in the PbS as well as in the ZnO was included in the numerical analysis used.

We begin our examination of cooperative diffraction found in the diffraction of Figure 1.7 by considering a plane wave with a wave vector of magnitude k_0 impinging from air onto an interface, with a periodic array, as depicted in Figure 1.7. Impingement is taking place at an angle φ to the array normal. The elements of this array have a diameter d and a spacing L where $L \gg d$. As a result of this size disparity, the element's optical properties and diameter are of no importance in the analysis we employ. The wave scattered off the arbitrarily chosen left element in Figure 1.7 has a phase change $k_0 L \sin \theta$ at wavefront 3 compared to what it had at wavefront 1. On the other hand, the wave scattered off the next right element has a phase change $k_0 L \sin \varphi$ between wavefronts

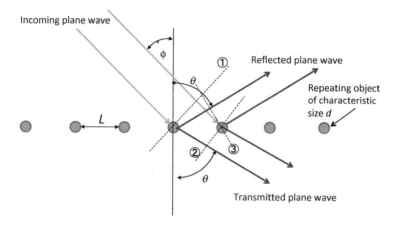

Fig. 1.7. An incoming plane wave impinging on a periodically structured interface at an angle φ with respect to the interface normal. A wavefront of this impinging wave is shown by ①. Possible diffracted waves (wavefront ② and corresponding transmitted wavefront ③) are shown diffracted at angles θ_m by the periodically structured interface. As explained in the text, constructive interference can occur resulting in diffraction for the back reflected and for the transmitted waves. Diffraction by this array has a profound effect on the impinging photon momentum.

1 and 3. For these waves to constructively interfere (i.e., for 3 to actually be a wavefront of a back-reflected plane wave), it is necessary that $k_0 L \sin \theta - k_0 L \sin \varphi = m2\pi$ or

$$k_0 \sin \theta_m = k_0 \sin \varphi + \frac{m2\pi}{L} \tag{1.5}$$

where $m = \pm 1, \pm 2, \dots$ If the incoming wave is impinging normally, then Eq. 1.5 reduces to

$$\sin \theta_m = \frac{m\lambda_0}{L} \tag{1.6}$$

In both equations, the θ_m generated by the $m = \pm 1, \pm 2,\dots$ possibilities are the angles that give back-directed diffraction into the air. Note that $m = 0$ in the general Eq. 1.5 and in the specific Eq. 1.6 formulations is the condition for specular reflection.

Of course, the incoming wave may also be diffracted and transmitted. In that case, diffraction will also produce wavefront 2 for the transmitted plane wave seen in Figure 1.7 for the diffraction angles θ_m given by Eqs. 1.5 or 1.6. This transmitted wave will undergo refraction if it is in a different material below the array. To account for refraction, the k_0 on the left-hand side of Eq. 1.5 must be replaced by nk_0 and the λ_0 in Eq. 1.6 must be replaced by λ_0/n to account for the transmitted wave entering a medium other than air (i.e., a material with $n > 1$) below the array.

While Eqs. 1.5 and 1.6 predict the diffraction angles θ_m for the reflected and transmitted waves, they also have a very profound photon momentum interpretation. This arises from the fact that k_0 is the magnitude of momentum of a photon[8] associated with the incoming wave. Therefore, the momentum $k_\parallel$ parallel to the interface before scattering is given by $k_\parallel = k_0 \sin\varphi$. It also follows from Eq. 1.5 that after scattering, $k_\parallel$ parallel to the interface is now given by

$$k_\parallel = k_0 \sin\varphi + \frac{m2\pi}{L} \qquad (1.7)$$

with $m = \pm1, \pm2, \ldots$ In other words, each photon in the incoming wave suffers a "collision" with the array at the interface, which enhances the photon momentum parallel with the interface by the amount $\pm(m2\pi)/L$. Since we will be interested in the transmitted diffracted wave's involvement in light trapping as it crosses an interface into a new medium, we need to account for refraction. It follows from the discussion above that Eq. 1.7 is generalized to Eq. 1.8

$$k_\parallel = nk_0 \sin\theta = k_0 \sin\varphi + \frac{m2\pi}{L} \qquad (1.8)$$

to account for diffraction and transmission into a medium with $n > 1$. Here again $m = \pm1, \pm2, \ldots$

As we see, the collision of an incoming photon with a one dimensional (1-D) array of periodicity L changes the photon's $k_\parallel$ momentum component by integer multiples of $2\pi/L$ where we will term $2\pi/L$ the reciprocal lattice vector. Here we are using the standard solid-state physics terminology that $2\pi/L$ is the 1-D reciprocal lattice vector of a simple 1-D reciprocal lattice defined by the real-space periodic array of Figure 1.7 [16]. From this perspective, the $\pm(m2\pi)/L$ are lattice points in

---

[8] Really $\hbar k_0$ but everyone tends to be lazy about bothering to multiplying by $\hbar$.

the array's 1-D reciprocal-space or k-space, as it is also called [16]. These $\pm(m2\pi)/L$ are also the lattice momenta, which can be imparted to the photons by the periodic array of Figure 1.7. Equation 1.8 shows that diffraction can even give a $k_\parallel$ component to a photon that is initially perpendicularly impacting the interface of the device! It is important to note that these reciprocal lattice vectors, which can do so much are determined solely by the array lattice and are independent of the impingement angle φ. The ability of a periodic array to turn wave vectors (and photon momentum) of incoming light so they have $k_\parallel$ components, or so they can have a variety of allowed $k_\parallel$ components parallel to the interfaces of the absorber, is of considerable interest. As we will see in Chapter 3, this phenomenon can be useful in allowing light to enter into the modes present in a device.

While we developed our ideas here using the 1-D array of Figure 1.7, we anticipate the ability of lattice diffraction coming from any periodic array of surface features to be useful in light trapping. Figure 1.8 gives an example of a periodic array of domes. This two-dimensional (2-D) array has the reciprocal 2-D k-space lattice as seen in Figure 1.9.

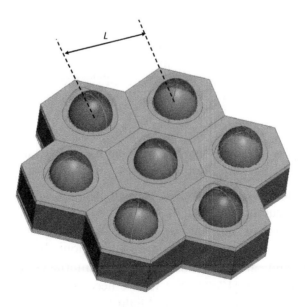

Fig. 1.8. A 2-D real-space lattice of hexagonally arranged domes.

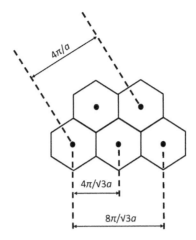

Fig. 1.9. The 2-D reciprocal lattice (k-space) corresponding to the 2-D periodic array of Figure 1.8. (Here L is the length of one edge of the hexagon.)

The lattice points of Figure 1.9 are 2-D analogs to $2\pi m/L$ and represent examples of the various momenta this 2-D real-space lattice can impart to impinging photons.

1.5 PLASMONICS

Incoming E-M radiation can cause plasma oscillations of the huge number of free electrons in any metal present. These are oscillations of the electrons about the positively charged cores of the metal [1]. Such plasma oscillations are quantized and the quanta are referred to as plasmons. For example, if we consider a metal mirror interface to a dielectric, or in the case of Figure 1.5 to an absorber, localized surface plasmon resonances (LSPRs) can arise [17–20]. These plasma oscillations can reradiate for certain incoming frequencies. Their use has been advocated as a way to advantageously modify scattering at the front and reflection at the back of a structure, such as that of Figure 1.5 [17–22]. While these plasma oscillations in the metal also dissipate energy due to collisions of electrons within the plasma, the use of plasmonic behavior at front grids/meshes and at back absorber/metal interfaces has received considerable attention [17–22], but its use necessitates considerable effort at controlling collisional loss and light blockage in front structures and collisional loss in back structures.

Single metallic nanoparticles in proximity to an absorber can sustain LSPRs and the near-field that exists around them can modify the absorption properties of the near-by material [18]. Two additional effects can come, interestingly, from a periodic array of nanostructures (nanoparticles or gratings) at a front structure or back absorber/metal interface [18]. These effects come from the fact that such periodic arrays can set up collective resonances resulting from the cooperative, radiative coupling of LSPRs. The first effect consists of hybridization of the trapped traveling modes in thin dielectric or absorber guiding layers with the LSPRs of a nanostructured periodic metal array [18]. This hybridization is referred to as surface plasmon polaritions (SPPs). For these hybridized modes, a wave-guiding structure giving rise to trapped traveling modes is necessary. Such wave-guiding structures will be discussed in Chapter 2. The second mechanism relies on diffracted orders of the reradiated energy due to the periodic structure in the plane of the array. These can lead to the hybridized resonances known as surface lattice resonances (SLRs) [17–22]. Both of these cooperative, coupled modes can exist in the absorber and perhaps advantageously affect light trapping and absorption there.

1.6 REFRACTION

Refraction is listed as a possible contributor to light trapping in Table 1.1 and we will now take a look at this possibility with the help of Figure 1.10. Here a plane wave with the wave vector $\mathbf{k}_0$ is seen impinging on a curved air/TCO interface at a direction normal to the device substrate plane. The photons of this impinging wave have no momentum $k_\parallel$ parallel to this device substrate plane but, as was the case for diffraction, refraction can change this situation. If we follow the refraction of this wave, we see that the magnitude of its wave vector changes to $\mathbf{n}_T\mathbf{k}_0$ in the TCO layer and the momentum of the photons develops a no-zero $k_\parallel$ component. This situation further evolves for this wave in the absorber where it now has a wave vector magnitude of $\mathbf{n}_A\mathbf{k}_0$ and a $k_\parallel$ component given by

$$k_\parallel = \mathbf{n}_A\mathbf{k}_0\cos\left(90-\varphi-\beta+\gamma+\theta\right) \tag{1.9}$$

Here, $\gamma = \arcsin[(n_T\sin\beta)/n_A]$ and $\theta = \arcsin[(\sin\varphi)/n_T]$.

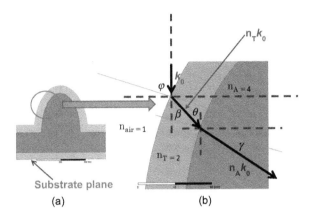

Fig. 1.10. A region of a dome-like structure (a) is enlarged (b) to aid in exploring refraction. Incoming plane wave is seen to be impinging on an air/TCO interface at an angle φ with respect to the interface normal at point of impingement. An absorber is seen to lie below the TCO layer.

As was the case with diffraction, our thought is that turning the wave vectors (and photon momentum) of incoming light so they have $k_{\parallel}$ components should act to keep the light in the device; i.e., it should be beneficial for light trapping. While we have seen diffraction can accomplish this turning of the wave vector (and photon momentum), we now see refraction can, also.

CHAPTER 2

Modes and Hybridization

2.1 INTRODUCTORY COMMENTS

Figure 2.1 shows that when light is present in a planar solar cell or detector structure, it is there in either radiation or trapped traveling modes[1] [7]. However, the trapped traveling modes are moving laterally and are inaccessible to incoming radiation unless there is some event that changes the incoming photon momentum. Such an event can be scattering arising from textured interfaces, for example, or diffraction. We will discuss these much more in Chapter 3. We also will soon see that light can exist in additional modes in nanoelement[2] array architectures. We classify these additional modes as trapped localized modes and hybridized modes.

2.2 RADIATION MODES

Let us start our discussion with a look at the radiation modes of Figure 2.1. If light enters into this planar structure of Figure 2.1, it can do so only by refraction. Consequently, impinging light must refract into the radiation states cone defined by $\sin \theta = n^{-1}$ in the figure. We already took a brief look at the radiation modes of Figure 2.1, when we considered normal impingement of a plane wave onto the planar structure shown in Figure 1.5. We utilized a numerical solution of Maxwell's equations to determine the reflection loss at the top control surface of Figure 1.1. The results are given as a function of wavelength in the plots of Figure 1.6. We saw that at some wavelengths this reflection back out the top control surface can be very small resulting in effective trapping of photons and their conversion in the absorber into charge carriers or excitons. On the

---

[1] As is our practice, the word "modes" is used interchangeably with the terms "states" and "wave functions."

[2] The terms nanoelement and nanoscale are being invoked liberally. Their use in this text encompasses structures that may be several hundred nanometers in size.

Introduction to Light Trapping in Solar Cell and Photo-detector Devices. DOI: 10.1016/B978-0-12-416649-3.00002-6

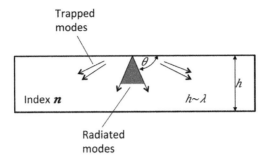

Fig. 2.1. Light in the trapped and radiation modes of a planar structure. The trapped traveling modes are discrete, as seen in the figure when λ ~ h but form a continuum when λ < h. (Adapted from Ref. [7].)

other hand, we saw that at some wavelengths this reflection loss can be very large giving Fabry–Perot reflection peaks and significant light energy leaving at the control surface of Figure 1.1.

We now take a closer look at the role of radiation modes by using an analytical treatment of this reflection situation we previously studied numerically in Chapter 1. Using the conventions of Figure 2.2 to facilitate our analytical approach, we see the analysis can be 1-D and couched in terms of a plane wave $e^{i(kx-\omega t)}$ crossing the control surface from the

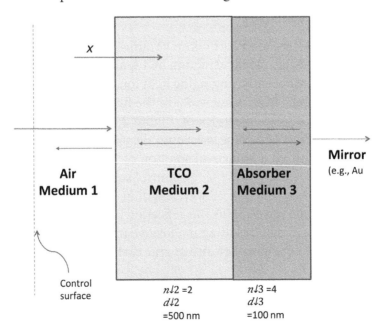

Fig. 2.2. A convenient version of Figure 1.5 for our analytical analysis.

left onto the structure. This wave enters the device and a plane wave $Re^{-i(kx+\omega t)}$ bounces back across the control surface to the left, away from the structure. For our analysis let us assume that this impinging wave has its electric field $\vec{\xi}$ oriented parallel to the layers of the structure and has an intensity normalized to unity. The loss $Re^{-i(kx+\omega t)}$ escaping from the structure is produced by a multiplicity of reflections and transmissions taking place inside the structure of Figure 2.2. We can write the net result of all these interactions as two oppositely moving radiation modes in layer 1 given by $A_1 e^{ik_1 x} e^{-i\omega t} + B_1 e^{-ik_1 x} e^{-i\omega t}$ and two oppositely moving radiation modes in layer 2 given by $A_2 e^{ik_2 x} e^{-i\omega t} + B_2 e^{-ik_2 x} e^{-i\omega t}$. There is only an evanescent wave of the form $Te^{-\epsilon x} e^{-i\omega t}$ in the metal.

Using these solutions for each region and Stokes' theorem, it immediately follows from the assumed electric field polarization that [23]:

$$1 + R = A_1 + B_1 \tag{2.1a}$$

at the air-1 interface ($x = 0$);

$$A_1 e^{ik_1 d_1} + B_1 e^{-ik_1 d_1} = A_2 e^{ik_2 d_1} + B_2 e^{-ik_2 d_1} \tag{2.1b}$$

at the 1-2 interface ($x = d_1$); and

$$A_2 e^{ik_2(d_1+d_2)} + B_2 e^{-ik_2(d_1+d_2)} = Te^{-\epsilon(d_1+d_2)} \tag{2.1c}$$

at the 2-metal interface ($x = d_1 + d_2$). Equation 2.1c conveys the fact that the excitation in the metal is evanescent and of the form $Te^{-\epsilon x}$ where ϵ is the reciprocal of the penetration depth. Each of these 2.1 equations has been divided by $e^{-i\omega t}$. Three additional relationships involving the A, B, and T quantities are required to give a determined mathematical system for obtaining these six coefficients. These come from Fresnel equations [23–26] and are discussed in Appendix B. The various k quantities in Eqs. 2.1 are related to the incoming k by $\tilde{n}_1 k = k_1$, $\tilde{n}_2 k = k_2$, and $\tilde{n}_3 k = k_3$. Here $\tilde{n}_j$ is the complex index of refraction discussed in Appendix C. This complex index of refraction is, in general, dispersive; i.e., each component can be a function of frequency.

As discussed in Appendix B, the form of these relationships, as well as those of Eqs. 2.1, vary with polarization. Solving this set of six equations

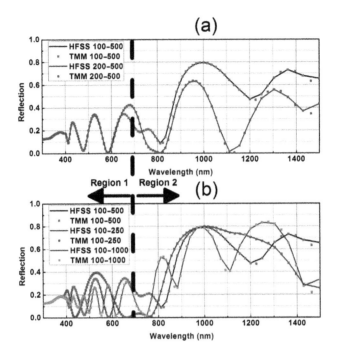

Fig. 2.3. Plots of $R^2(\lambda)$ obtained from analytical (TMM) and Maxwell's equations numerical solver (HFSS) results for Figures 1.5 and 2.2 for (a) fixed TCO (AZO) and varying absorber (PbS) thicknesses and (b) varying TCO (AZO) and fixed absorber (PbS) thicknesses. Experimental $\tilde{n} = n(\omega) + in_i(\omega)$ data were used. First number in the legend is the PbS thickness; the second is the AZO thickness.

and six unknowns allows the function $R(\lambda)$ to be obtained. The transfer-matrix method (TMM) solution methodology was used here for analytically solving the set and finding $R_A(\lambda)$ [25]. Once $R(\lambda)$ is determined, the reflection as a function of wavelength follows from $R^2(\lambda)$.

Figure 2.3 shows plots of $R^2(\lambda)$ as a function of (air) wavelength as obtained by both our analytical analysis (TMM) and by numerical solutions of Maxwell's equations (HFSS). As can be seen, the results coming from using the analytical model we constructed match the exact results obtained by numerically solving Maxwell's equations. This agreement supports several points: (1) our physical interpretation of how the radiation modes are involved is essentially correct and (2) our mathematical representation of these radiation modes in the structure of Figure 1.5 (Figure 2.2) is essentially correct.

Figure 2.4 shows plots of $R^2(\lambda)$ for which the TCO thickness has been fixed at 60 nm but the absorber thickness has been varied. Importantly,

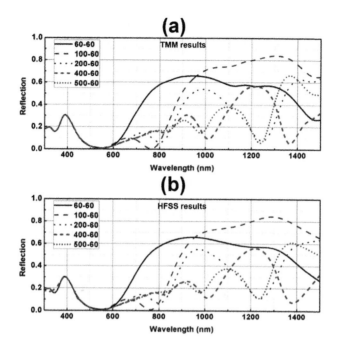

Fig. 2.4. Examining the ¼ wavelength standard ARC coating approach using plots of $R^2(\lambda)$ obtained from analytical (TMM) and Maxwell's equation solver numerical (HFSS) results. Here the ARC coating is fixed at a 60-nm thickness and the PbS absorber thickness is taken as 60, 100, 200, 400, and 500 nm. Experimental $\tilde{n} = n(\omega) + in_1(\omega)$ data were used. First number in the legend is the PbS thickness; the second is the AZO thickness.

the TMM and HFSS data are again seen to agree well giving further support to our picture of the role of radiation states for this case of a plane wave impinging on a planar structure. The 60-nm value has been picked here for the top layer since it is roughly a quarter of the wavelength of light in the TCO (where $n \cong 2$) at the mid-range of the 300–700 nm (near UV to visible) part of the solar spectrum. Such quarter wavelength, wide band gap materials with $n \cong 2$ are effective simple antireflection coating (ARC) structures for solar cells and photodetectors [26–28]. Consistent with that design rule, Figure 2.4 demonstrates the 60-nm thickness for the $n \cong 2$ top layer is very effective in suppressing reflection for most of ~300–700-nm range for absorber thickness >ARC thickness cases.[3] Interestingly, this design rule begins to fail for absorber thickness ≈ARC

---

[3] An ARC coating can reduce the reflection to an average of 10% consistent with Figure 2.4. A two-layer ARC coating can reduce the reflection to an average of 3% [28].

thickness. This ratio ≈ 1 situation is commonly found in today's thin film solar cells as well as in thin photodetector structures. As can be inferred from our discussion of the interference in Chapter 1, in this thickness ratio regime, reflection from the whole materials system must be considered for thin structures as is correctly done in Figure 2.4.

2.2.1 Dispersion Relations for Radiation Modes

The expressions used in the proceeding analysis connecting the k in each material with that of the incoming wave in air give a general relationship between a wave's propagation vector in a medium and its frequency. This is termed a dispersion relation. While dispersion relationships[4] vary with material and situation, they all have the form $\omega = \omega\left(\vec{k}\right)$ when thinking in terms of waves and the form $E = E\left(\vec{k}\right)$ when thinking in terms of photons. In general, they may be single valued or multivalued; they may be in a continuum or bands. For radiation modes in a homogeneous medium, they are simply of the form

$$\omega = \left(\frac{c}{n}\right)\left|\vec{k}\right| \tag{2.2a}$$

Correspondingly, the relationship between E and $\vec{k}$ for the photons associated with this wave is

$$E = \left(\frac{\hbar c}{n}\right)\left|\vec{k}\right| \tag{2.2b}$$

Equations 2.2a and b may also be written as

$$\omega = \left(\frac{c}{n}\right)\sqrt[2]{k_x^2 + k_y^2 + k_z^2} \tag{2.3a}$$

and

$$E = \left(\frac{\hbar c}{n}\right)\sqrt[2]{k_x^2 + k_y^2 + k_z^2} \tag{2.3b}$$

---

[4] Also called frequency or photon-energy bands or band diagrams.

In these expressions n is the real part of the index of refraction of a medium (see Appendix C).

The easiest way to plot $\omega = \omega(\vec{k})$ or the corresponding $E = E(\vec{k})$ in three dimensions (3-D) is to show $\omega = \text{constant}_1$ or $E = \text{constant}_2$ surfaces in a 3-D k-space. In two dimensions (2-D), the corresponding plots have $\omega = \text{constant}_1$ or $E = \text{constant}_2$ curves in a 2-D k-space. In one dimension (1-D), things get really easy and one can simply plot E or ω versus k. These possibilities are seen in Figure 2.5.

We will now assume that the various media our waves are traveling in (except metals) are loss-less. We do this to easily use insightful ideas and approaches from dielectric waveguide theory. We will still lable regions as TCO, absorber, etc., but to get started, we will think of them as loss-less. The dispersion relationships of loss-less media tell us any wave of frequency ω is a traveling wave so long as its $\vec{k}$ components k_x, k_y, and k_z seen in Eqs. 2.3a and b are all real numbers. On the other hand, Eqs. 2.3a and b also tell us that one can pick a combination of (k_x, k_y, k_z) such that one or more can be imaginary numbers. In this situation, the wave must be evanescent; i.e., decaying.

We now introduce the quantity $\vec{k}_{\parallel}$ used in dielectric waveguide discussions [24] and seen in Figure 2.6 by choosing to write any wave vector $\vec{k}$ as $\vec{k} = \vec{k}_{\parallel} + \vec{k}_z$ where $\vec{k}_{\parallel} = \vec{k}_x + \vec{k}_y$ lies in the $k_x - k_y$ plane of Figure 2.7. This $\vec{k}_{\parallel}$ choice may seem capricious but it is not. It will prove very useful, since, in some future discussions, we will take the $k_x - k_y$ plane to be parallel to the light impingement surface. This use of $\vec{k}_{\parallel} = \vec{k}_x + \vec{k}_y$ allows Eqs. 2.3a and b to be rewritten as

$$\omega = \left(\frac{c}{n}\right)\sqrt[2]{k_{\parallel}^2 + k_z^2} \tag{2.4a}$$

and

$$E = \left(\frac{\hbar c}{n}\right)\sqrt[2]{k_{\parallel}^2 + k_z^2} \tag{2.4b}$$

From Eq. 2.4a, we see that we can think of the cone in Figure 2.5c or in Figure 2.6 as being a plot of this equation with $\vec{k}_z = 0$ (for the $n = 1$ case). With this in mind, we have denoted the abscissa in Figure 2.6 as

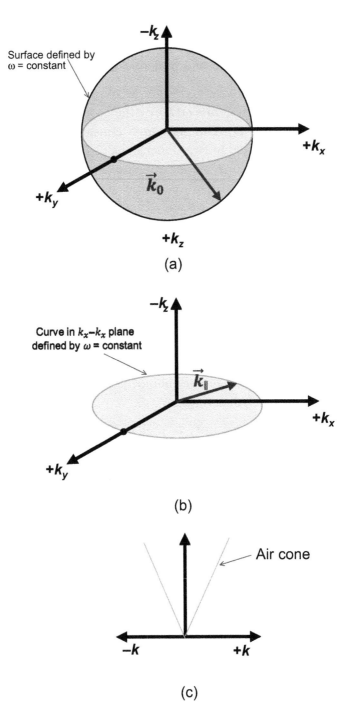

Fig. 2.5. (a) A constant frequency surface depicted in 3-D k-space for the dispersion relationship of Eq. 2.3a. Such surfaces are spheres even if n is a function of frequency. (b) The 2-D plot corresponding to the intersection of a surface of (a) with the $k_x - k_y$ plane. The resulting curve (a circle even if n is function of frequency) describes waves with $\vec{k}$ having only k_x and k_y components. (c) A 1-D plot giving the frequency values encountered following a radius k in (a) or (b).

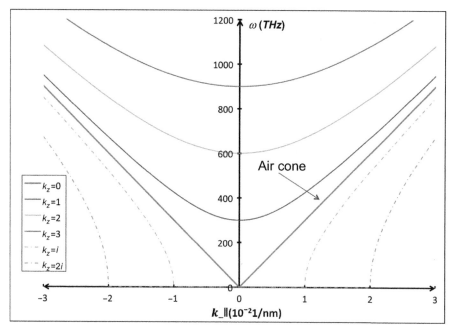

Fig. 2.6. The continuum of dispersion relations $\omega = \omega\,(k_\parallel)$ that occur by taking k_z^2 as a parameter. Three traveling wave and two evanescent wave examples are sketched in this "projected band diagram."

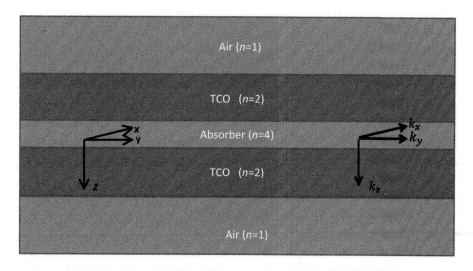

Fig. 2.7. An example of a waveguide structure. The guided mode depends on z and not y. It only propagates in the $k_\parallel$ direction where $k_\parallel$ lies in the $k_x - k_y$ plane.

$k_{\parallel}$. This allows us to use Figure 2.6 and Eq. 2.4a to realize that, for a given value of $k_{\parallel}$, there is a continuum of traveling wave states (ω, k_z) available in a medium with ω values lying above the cone at the given $k_{\parallel}$. Figure 2.6 and Eq. 2.4a also allow us to note that for the same given value of $k_{\parallel}$, there is a continuum of evanescent wave states (ω, k_z) available in the medium with ω values lying below the cone at the given $k_{\parallel}$. Another way to sum up what we have learned is to say that there is a continuum of dispersion relations $\omega = \omega(k_{\parallel})$, which can be generated by treating k_z^2 as a parameter in Eq. 2.4a. One evanescent and three traveling wave dispersion relations of this type are shown as examples in Figure 2.6. While we couched these observations in terms of Eq. 2.4a, the same conclusions follow for Eq. 2.4b.

Plots such as Figure 2.6 are sometimes called projected band diagrams [24]. Working with these $\omega = \omega(k_{\parallel})$ or equivalently $E = E(k_{\parallel})$ radiation mode dispersion relationships will prove quite helpful in some of our discussions of light trapping since we will take the normal to the $k_x - k_y$ plane to be parallel to normally impinging light. Any wave in the structure that is propagating with only the vector $\overrightarrow{k_{\parallel}}$ would be moving parallel to this top surface, an interesting situation for light trapping in the absorber.

When a wave moves from one material to another, the applicable dispersion relationship must change. If we couch this situation specifically in terms of photons, a photon's moving from one material to another means that, in k-space, it moves from an initial state $\left(E_i, \overrightarrow{k_i}\right)$ in the initial material to a final state $\left(E_f, \overrightarrow{k_f}\right)$ in the final material. When this happens, there are certain rules that must be followed in determining the new state. First, the photon energy E_i or, equivalently, the wave angular frequency ω_I must be conserved. Second, the component of the wave vector $\overrightarrow{k}$ that is parallel to the interface between material I and material f must be conserved or, in the language of photons, the photon momentum parallel to the interface must be conserved [24]. We already saw these rules at work in Chapter 1 where they are obeyed for reflection (Snell's law), for refraction, and to within ±integer multiples of the lattice momentum, for diffraction. We interpreted this last observation by saying that photon collisions with a periodic interface array give the photons ±integer multiples of the lattice momentum parallel to the interface.

2.3 TRAPPED TRAVELING MODES: GUIDED MODES

Let us consider now trapped traveling modes of the type indicated in Figure 2.1. To understand their origin in Figure 2.1, we need to think of the structure as being a laterally featureless waveguide [17,23–26].

2.3.1 Dispersion Relations for Guided Modes

The behavior of $\omega = \omega(\vec{k})$ or equivalently $E = E(\vec{k})$ for a guided wave must often be obtained in numerical form from numerical solutions to Maxwell's equations and the applicable boundary conditions [24]. However, we can pursue exploring waveguide effects by continuing to use loss-less materials. With this approach, a more concrete, detailed example of the device sketched out in Figure 2.1 may be found in the stack configuration depicted in Figure 2.7. The structure is seen to consist of a high index of refraction (e.g., $n = 4$) absorber layer embedded between two layers of a TCO material with a lower index of refraction (e.g., $n = 2$). As noted, for now we will continue to think of the "absorber" as being loss-less. The overall structure is surrounded by air ($n = 1$). While this is not even a simple solar cell or detector structure, it is an example of a waveguide and easily seen to be very similar to a solar cell or photodetector in its configuration. A wave launched parallel to the lower index/higher index interfaces will congregate in the high index material where its propagation will be guided by the surrounding lower index medium [17,24]. The waves existing in the embedded, high n absorber layer are what we have called guided modes. Their electromagnetic field ξ resides mostly within the high n layer and their propagation is controlled solely by the vector $\vec{k}_{\parallel}$, which lies in the x–y plane parallel to the absorber layer interfaces. The functional form of these guided modes must follow [24]:

$$H_{k_{\parallel}}(\vec{r}) = h(z)e^{ik_{\parallel}\rho} \tag{2.5}$$

where $k_{\parallel\rho}$ is the result of the dot product of the $\vec{k}_{\parallel}$ vector and the position vector $\vec{\rho}$ in the x–y plane. Because $\vec{k}$ only has the $\vec{k}_{\parallel}$ component lying in the $k_x - k_y$ plane, the dispersion relationship for these guided modes must have the form $\omega = \omega(k_{\parallel})$. This point underscores why we anticipated that the form of Eqs. 2.4a and b and plots such as those of Figure 2.3 would be useful.

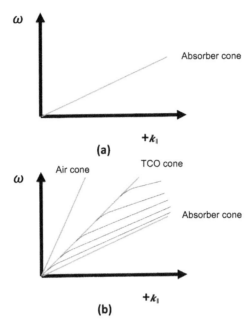

Fig. 2.8. (a) The function $\omega = \omega(k_{\parallel})$ for an unconfined, isotropic, homogeneous absorber. The absorber light cone only is shown. (b) A sketch showing the function $\omega = (k_{\parallel})$ for the absorber when placed in the structure of Fig. 2.7. Here the continuum of $\omega = \omega(k_{\parallel})$ possibilities for the absorber for (a) has condensed into bands due to confinement.

Let us suppose for a minute that the absorber material of Figure 2.7 was not confined and instead is of infinite extent, then its projected band structures $\omega = \omega(k_{\parallel})$, which we get from Eq. 2.4a by taking varying k_z^2 as a parameter, would look like the sketch in Figure 2.8a. Figure 2.8a shows the absorber's light cone in the $\omega = \omega(k_{\parallel})$ dispersion relationship space. All the states above the cone would be traveling waves in the nonabsorber medium and all those below the cone would be evanescent waves in the nonabsorber medium. If we now place this absorber back into the structure of Figure 2.7, Figure 2.8a changes dramatically to Figure 2.8b. There is no longer a continuum of $\omega = \omega(k_{\parallel})$ dispersion relationships for the absorber. In fact, the dispersion relationship for the absorber may now be bands, as sketched in Figure 2.8b. These will correspond to different $h(z)$ of the guided wave functional form given by Eq. 2.5. The allowed dispersion relations will coalesce onto ω (or E) curves with diminishing waveguide height in the z-direction. This behavior is analogous to the increasing difference between electron energy levels in quantum dots (QDs) with decreasing QD diameter [1].

In the case of electrons, this effect is termed quantum confinement. The fact that the bands of Figure 2.8b become a continuum with increasing waveguide height is analogous to the continuous energy levels available to electrons in a bulk solid.

The change from Figure 2.8a to Figure 2.8b can be summarized by saying confinement causes the continuum of $\omega = \omega(k_{\parallel})$ possibilities in the absorber to condense into guided wave bands. These bands cannot extend to the right of the absorber cone since Eq. 2.4a shows k_z^2 would be negative forcing the waves to be evanescent. One further comment is needed: since the TCO is confined, its dispersion relationship has coalesced into bands too. This and the continuum of dispersion relations for the air region are not shown in Figure 2.8b for clarity.

2.4 TRAPPED TRAVELING MODES: BLOCH MODES

Let us now consider the type of trapped traveling modes termed Bloch waves. These do not arise in a simple waveguide structure so they are not among those depicted in Figure 2.1. They require a photonic crystal structure; i.e., they require absorber periodicity [24]. The presence of a periodicity can have a profound effect on the nature of the dispersion relation and the wave functions of the absorber. A very simple example of a periodic 2-D absorber configuration may be seen in Figure 2.9, which

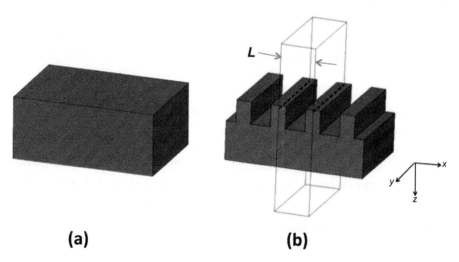

(a)   **(b)**

Fig. 2.9. (a) A nonperiodic absorber and (b) a 2-D version of this absorber with 1-D periodicity of period L. The absorber is surrounded by air in (b). These structures are of infinite extent in the y-direction.

gives an absorber in a nonperiodic configuration (Figure 2.9a) and in a periodic configuration (Figure 2.9b). As seen, the absorber shows 1-D periodicity in the x-direction[5] with period L in Figure 2.9b; i.e., the configuration has a unit cell of width L. This periodicity forces the E-M waves in the absorber of Figure 2.9b to obey Bloch's theorem; i.e., they must be of the form [24]

$$H_{k_\parallel}(\vec{r}) = u_{k_x}(x,z)e^{ik_x x} \tag{2.6}$$

where the function $u_{k_x}(x,z)$ modulating the plane wave has to obey $u_{k_x}(x,z) = u_{k_x}(x+mL,z)$ with $m = \pm 1, \pm 2, \pm 3,...$ [24]. A wave function with the properties of Eq. 2.6 is called a Bloch state, Bloch mode or Bloch wave. A key property of the dispersion relation $\omega = \omega(k_x)$ for the Bloch waves of Eq. 2.6 is that the function is periodic in k-space [24]:

$$\omega(k_x) = \omega\left(k_x + l\frac{2\pi}{L}\right) \tag{2.7}$$

With $l = \pm 1, \pm 2, \pm 3,...$ In other words, $\omega = \omega(k_x)$ must have periodicity of $2\pi / L$. It may also be noted from Eq. 2.7 that the momentum component k_x has lost its precise meaning. Now, k_x and $k_x + l(2\pi / L)$ are the same; i.e., Eqs. 2.6 and 2.7 show they give the same wave function, propagation direction and frequency (or photon energy). While at first this all may look like what we saw for diffraction in Chapter 1, it is really very different. In the case of diffraction at a surface array, that changes the wave function, its propagation direction, and photon momentum.

2.4.1 Dispersion Relations for Bloch Modes

The points $l(2\pi / L)$ $(l = \pm 1, \pm 2, \pm 3)$ can be utilized to construct a 1-D k (reciprocal) space lattice with unit cell width of $2\pi / L$, which is convenient because we can plot the dispersion relation $\omega(k_x)$ in this reciprocal space. Since $\omega(k_x)$ is an even and periodic function, Eq. 2.7 tells us that all the information about $\omega(k_x)$ can be obtained by plotting the function between $-(\pi / L)$ and $+(\pi / L)$ of this reciprocal space. The

---

[5] The notation is switched from $k_\parallel$ to k_x here since k_x is in a specific direction: that of the 1-D periodicity.

whole $\omega(k_x)$ function must fold into the $-(\pi/L)$ and $+(\pi/L)$ region due to its periodicity. This region of reciprocal space is called the first Brillouin zone or often it is simply called the Brillouin zone [16,24]. A depiction of the impact of the periodic dispersion relationship requirement forced on a periodically configured absorber is sketched in Figure 2.10. Here Figure 2.10a shows $\omega(k_x)$ for absorber when it is not periodically

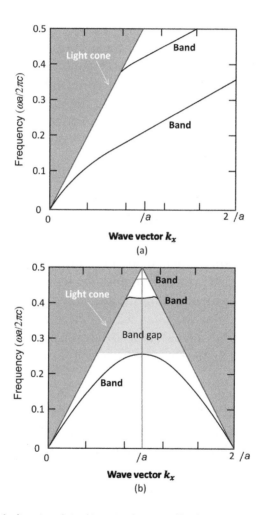

Fig. 2.10. Sketches of the dispersion relationships giving the projected band structure for a 2-D absorber surrounded by air for (a) when the absorber is not periodically configured and for (b) when the absorber is 1-D periodically configured with period a. Shading shows the region in this projected band structure k_x-space having a continuum of allowed states. In b, $k_x = (\pi/a)$ is the first Brillouin zone positive direction boundary and the region $-(\pi/a) < k_x < (\pi/a)$ is the first Brillouin zone. These values of k_x have no significance in part (a) due to lack of periodicity. Based on Ref. [24].

configured and Figure 2.10b shows $\omega(k_x)$ for the periodically configured absorber in air. Here the unit cell dimension is taken to be a.

Figure 2.10a and b are sketches but they convey the implications of absorber periodicity for the absorber dispersion relationship. The positive k-direction Brillouin zone boundaries of the first Brillouin zone and the second Brillouin zone are present in Figure 2.10b.[6] Interestingly, Figure 2.10b shows that the periodicity causes ranges where there are no allowed frequencies (i.e., these photon energies are not permitted to exist in our periodic absorber). This is the behavior of a photonic crystal [24]. These "band gap" regions are analogous to those found for the dispersion relationships for electrons in semiconductors. The origin is the same in both cases: the periodicity in the physical world in which the corresponding wave exists. One additional note: clearly solar cell and detector structures with 2-D absorber periodicity are possible. For example, this can be achieved with the structure of Figure 1.2b, c, and e by simply making the features absorber material and not TCO.

2.5 TRAPPED LOCALIZED MODES: MIE MODES AND PLASMA MODES

We have seen that a simple planar solar cell or detector will have radiation modes and trapped traveling modes for light trapping. The latter will be guided modes arising from the structure functioning as a waveguide. If the structure of the solar cell or detector is developed further to the point where it has a periodic absorber, the trapped traveling waves can be Bloch modes. If the device structure is further developed so that is has dielectric or semiconductor nanoscale features, it can then have trapped localized modes known as Mie modes. As we have noted earlier, these have their electric field localized around or in a wavelength-scale feature. If the nanoscale features are metals, then plasmonic modes can occur. In addition, these various radiation and trapped modes may interact giving rise to hybridization; i.e., the mixing of modes with the same photon energy. Interestingly, hybridization of trapped localized modes with radiation or trapped traveling modes requires nanoscale dielectric or metal topological or morphological features.

---

[6] A concise summary of how to form Brillouin zones in general 2-D and 3-D situations is found in Appendix A of Ref. [24].

CHAPTER 3

Light-Trapping Structures

3.1 INTRODUCTION

When light impinges on the entry (top) surface of a structure, the net result is some combination of reflection, refraction, and diffraction. Since we want light to enter inside the structure and become trapped, only refraction and diffraction at this surface are useful. Once the light is inside the structure, we can conclude from our discussions in Chapters 1 and 2 that there are, in our classification scheme, four types of modes in which photons can reside: (1) radiated modes, (2) trapped guided and Bloch traveling modes, (3) trapped, spatially localized modes (e.g., Mie modes and LSPRs), and (4) hybridizations of these. In this chapter, we explore and compare structures that employ various combinations of these light-trapping modes.

3.2 PLANAR STRUCTURES WITH ARCs

As we saw in Chapters 1 and 2, a planar device structure having a non-absorbing coating (i.e., an ARC) with a judiciously chosen thickness and index of refraction n (or graded n) can create internal radiation mode patterns that suppress entry-surface back reflection and loss. We saw properly designed ARCs could suppress Fabry–Perot reflection peaks over a wavelength range determined by the index of refraction variation and ARC thickness selections. Interestingly, an index of refraction change (abrupt or graded) does not necessarily have to be due to material variations with thickness. It can be due to nanofeatures on the surface of an absorber that are smaller than the wavelength of light in the range of interest (as an example, we picked the ~300–700 nm range back in Chapter 1). Such surface structures can give an effective index of refraction that varies with thickness due to the variation in the air/medium composition with thickness. Such an effective index of refraction is utilized in moths' eyes to reduce reflection; hence, the insect's name is given to this type of ARC reflection suppression [27].

Introduction to Light Trapping in Solar Cell and Photo-detector Devices. DOI: 10.1016/B978-0-12-416649-3.00003-8

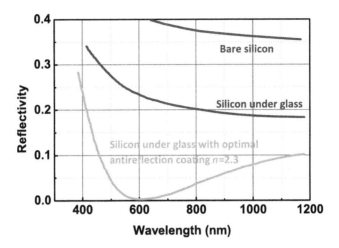

Fig. 3.1. Reflectivity as a function of (air) wavelength λ for bare Si, Si with cover glass, and Si with cover glass and ARC. Based on data taken from http://www.pveducation.org/pvcdrom/design/anti-reflection-coatings.

Figure 3.1 shows reflectance (reflection/incoming intensity) as a function of wavelength for bare Si single crystal material, Si single crystal material with cover glass, and Si single crystal material with cover glass and an ARC [28]. These data are for relatively thick Si absorber material where one can think of the situation as an ARC coating on a thick absorber. In the case of thin devices, which are our main interest, light can "see" the whole solar cell or photodetector as an effective AR materials system. In the case of detectors, such a structure is termed a resonant cavity-enhanced (RCE) photodetector [29]. Whether dealing with thick or thin devices, photovoltaic devices or photodetectors, the use of ARCs only involves trying to manipulate the radiated modes of Figure 2.1 and therefore its effectiveness is limited by their density of states. The ARC approach also is impingement angle dependent.

3.3 PLANAR STRUCTURES WITH RANDOMLY TEXTURED SURFACES

Structures with randomly textured[1] surfaces, interfaces, or some combination thereof have been used for many years for trapping light [1]. As we discussed in Chapter 1 and Appendix A, these structures are designed to

---

[1] Randomized surface topography and randomly textured surface are used synonymously in this text to describe a surface whose reflection is approaching, to some degree, a Lambertian distribution.

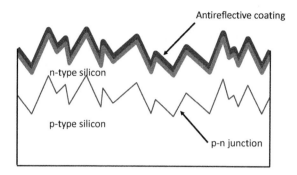

Fig. 3.2. An absorber (having a p-n junction) with texturing at the top surface and at the junction interface. An ARC is also present.

make effective use of multiple reflections with the objective of approaching complete internal reflection and randomization of the energy and propagation directions of light inside a structure. The multiple reflections "mix everything up" allowing the mode utilization to approach a Lambertion distribution. When only (1) radiated modes and (2) trapped moving modes are present, then both become equally involved [7]. One can say, from the perspective of geometrical optics, that by filling all these various modes due to multiple scattering events, the optical path length and therefore absorption is being increased. There are variations of this approach in which random texturing is present on the top surface, on the bottom surface, on intermediate interfaces, or on some combination of these. Some utilize an ARC in conjunction with the random texturing, as seen in Figure 3.2. The decision on where the random texturing resides depends on the materials, processing flow, and absorber thickness. Surface roughness is not the only way to achieve random scatters at a surface or interface; e.g., randomly positioned metal nanoparticles can be used too for plasmonic scattering. This is seen in Figure 3.3 [21,30]. However, as we have noted, scattering in not an elastic event in this case.

The use of randomly textured interfaces and surfaces has seen extensive application for various forms of silicon [31–35]. As an example, absorptance versus wavelength is presented in Figure 3.4 for the case of random texturing plus an AR coating for multicrystalline silicon. Two processing approaches to the random texturing were tried for the structures of this figure and, as seen by comparing with the control, these texturing results were successful. It must be noted, however, that this

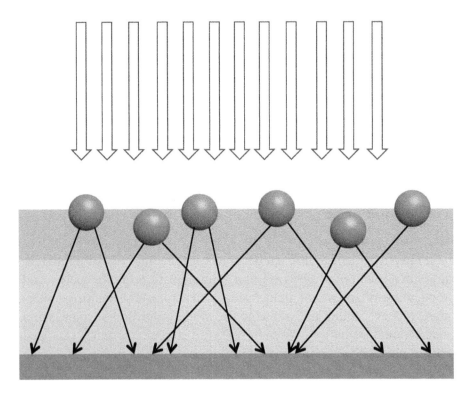

Fig. 3.3. Random scattering due to plasmonic scattering from randomly positioned metal nanoparticles.

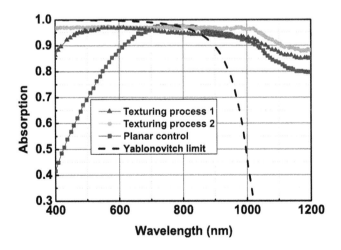

Fig. 3.4. Data for total absorptance versus (air) wavelength are shown for two random texturing processes and for a planar (nontextured) control. For comparison, the Yablonovitch limit (which gives absorption in the Si absorber only) is also shown. In all cases, the absorber in multicrystalline Si. Based on data take from Ref. [35].

plot shows total absorption and not just absorption in the Si absorber material. As a consequence, the three device curves for $\lambda > 1127$ nm (the Si band gap is 1.1 eV) are entirely due to loss in the other device materials. Even at shorter wavelengths, these losses are playing a significant role, as can be judged from the Yablonovitch Si absorption limit, which is expected to lie above the multicrystalline Si absorption data. While texturing can obviously improve light trapping, feature sizes of the texturing can be of the order of or larger than absorber film thicknesses. This can lead to shunting issues. In addition, irregular random texturing features may affect charge collection in thin films. Finally, in production a controlled surface topology randomization process step is required which can be viewed as a contradiction in terms.

3.4 STRUCTURES WITH NANOELEMENT ARRAYS

Randomly textured surfaces strive to attain excellent light trapping in planar structures by evenly populating photons into the available radiation and trapped moving modes. Of course, the ideal filling of these states with photons is achieved in the Yablonovitch limit. In this section, we begin our first explorations to determine if we can fill radiation and trapped moving states using fully controlled, not random, reproducible surface architectures. Importantly, we will also ask if we can populate the trapped localized and hybridization modes, which can be present in some architectures. These have not been involved in light trapping so far in our discussions of this chapter. The potential involvement of more modes is particularly interesting because the more modes open to storage, the more photons that can be trapped and absorbed. The fully controlled, not random, reproducible surface architectures that we have in mind are exemplified by Figure 1.1b, c, and e. We will term these nanoelement array architectures or, at times, just nanoelement architectures.

3.4.1 Discussion

Both the planar solar cell in Figure 3.5a and the nanoelement array solar cell in Figure 3.5b–c have radiation modes and trapped moving waveguide modes. Although Figure 2.8b is strictly speaking applicable to a loss-less "absorber", we have used its insight to determine that the trapped moving modes of a simple planar structure are not available to incoming photons. This is because, as Figure 2.8b shows, it would

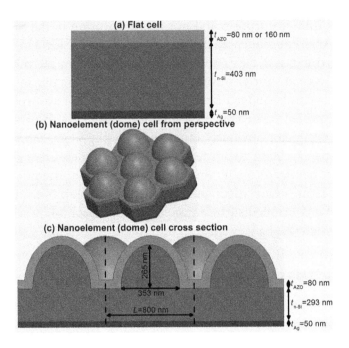

Fig. 3.5. (a) Conventional planar solar cell architecture, (b) a nanoelement array architecture in perspective, and (c) this nanoelement array architecture in cross-section. In this example, the nanoelements are domes. Dimensions appear on the figure.

be necessary for these incoming photons to have $k_\parallel$ values, for a given k_0 and therefore ω, that lie past the air and TCO light cone boarders in $k_\parallel$-space to occupy these trapped, moving waveguide states. To do so would require imaginary k_z components for waves traveling in air and TCO. Put succinctly, there are no radiation states having the required $k_\parallel$-values necessary to conserve both incoming photon energy and $k_\parallel$ and thereby qualify such photons to occupy these trapped, moving modes. So far, the only solution we have come up with that makes these trapped waveguide states available to incoming photons is to employ the scattering provided by randomly textured surfaces.

Actually, structures with the nanoelement array architecture exemplified in Figure 3.5b and c offer another solution that can make trapped, laterally moving waveguide states available to incoming photons for their ultimate absorption in the absorber medium. This solution is the involvement of diffraction that may play a role due to the presence of the periodic array. Very interestingly, we will see that the nanoelement array architecture presents other additional light-trapping possibilities including spatially localized states and hybridizations of these, which become available in these structures for incoming photons. Since architectures with features on this scale have these capabilities and are manufacturable with techniques such as nanoimprinting and nanomolding, they have become of considerable interest [36–40]. As seen in Figure 1.2b, c, and e, as well as Figure 3.5b and c, the hallmark of a periodic nanoelement array architecture is unit cells arranged in a lattice (hexagonal in Figure 1.2b, c, and e), which defines an array pattern. Depending on the nanoelements' relative height h versus the thickness t of the rest of the structure, Figure 3.5b and c may be viewed (1) as an array of nanoelements (which happen to be domes in this example) positioned on a 2-D waveguide with its (trapped moving) guided waves [41], (2) as a photonic crystal[2] with its (trapped moving) Bloch waves [24], or (3) as some intermediate.

As we have noted, the periodic nanoelement array arrangement of this architecture typified by Figure 3.5b and c can impart $k_\parallel$ additions to incoming photons. We demonstrated this photon momentum change due to photon "collisions" with a diffracting array for the simple case of a 1-D diffraction configuration in Chapter 1. Since these collisions do not change energy but only $k_\parallel$, some incoming photon (ω, $k_\parallel$) values now can be found past the air and TCO light cones in the $k_\parallel$-space sketch of Figure 2.8b due to diffraction. This means the range of radiation modes that may now be present in the absorber extends beyond the refraction limit $\sin \theta = n^{-1}$ of Figure 2.1. Generalizing Eq. 1.5, we see that diffraction extends the entry range for normally impinging light beyond $\sin \theta = n^{-1}$ to discrete angles θ_m given by $\sin \theta_m = n^{-1} + (m\lambda / nL)$ where $m = 1, 2, \ldots$ These additional radiation states are now available to be absorbed in the absorber medium thanks to diffraction. Some may have photon (ω, $k_\parallel$) values that fall on the dispersion relation for the absorber's trapped traveling modes. When this happens, an incoming, diffracted photon with such a (ω, $k_\parallel$) set can transition into the trapped, laterally moving waveguide mode with the same (ω, $k_\parallel$) set. This is seen in Figure 3.6 which uses the 1-D example of Chapter 1 and Figure 2.8b. Here the $k_\parallel$ values given to incoming photons by first, second, and third

---

[2] A photonic crystal is a periodic nanostructure that controls the flow of light. The E-M waves in such a structure are Bloch modes. These waves have the same periodicity as that of the structure. The guided modes of a wave guide have no periodicity requirement.

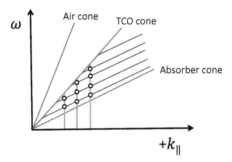

Fig. 3.6. The $k_\parallel$ values given to incoming photons by first, second, and third diffraction orders (vertical lines) intercept the trapped traveling modes at the circle labeling. The corresponding trapped traveling mode ω values can be populated by these incoming photons and can then be read from the ordinate.

diffraction orders, as examples, are shown outside the cone boarders. The resulting traveling mode ω values that can be populated by these photons are circled; the remaining possible ω values are radiation modes present due to the diffraction.

It is interesting to note that if we now assume light is impinging at the angle φ, use first order diffraction[3] as an example and continue with the simplified 1-D picture, then the first order $k_\parallel$ of Figure 3.6 is given by $k_\parallel = k_0 \sin\varphi + 2\pi / L$ from Chapter 1. If we vary the impingement angle φ in this expression, the corresponding vertical $k_\parallel$ line in the sketch of Figure 3.6 shifts and thus the allowed trapped traveling modes ω values shift. This says that light with various impingement angles can get into trapped modes but the value of ω depends on the impingement angle. Several groups have used the opportunity presented by this shift in ω and $k_\parallel$ with impingement angle φ to actually map the absorber's projected dispersion relation [42]. We have only sketched this dispersion relation in Figure 3.6.

In further evaluating the potential of nanoelement architectures such as the example portrayed in Figure 3.5b and c, it is interesting to visualize the surface of the nanoelements as being made-up of surface areas comparable in size to those of a textured surface. In a textured surface, each of these areas is randomly oriented. In Figure 3.5b and c each of the surface areas is systematically oriented with respect to the next in a manner dependent on the nanoelements' morphology. Looked upon in this manner, we

---

[3] The zeroth order diffraction into the device is really refraction and a radiation mode.

can say the nanoelement architecture refracts light into radiation modes and trapped modes in the structure over a wide variety of angles as is done by a textured surface. This refraction also depends on any material layers in the nanoelement, as seen in Figure 1.10. However, given the 3-D nature and morphology of the nanoelements of these architectures, the angular distribution of the refraction and subsequent reflection distribution produced is very biased and not random as it is for effective surface texturing. The latter, as we know, yields the Yablonovitch limit.

The nanoelement architecture of Figure 3.5 opens the door to still another type of photon states in a solar cell or photodetector structure. These are spatially localized modes and hybridizations of such spatially localized modes with other modes. The type of localized state that can be present when the nanoelements are dielectrics (here we are including absorber materials) is, as we have mentioned before, Mie modes [43]. These can be found associated with each nanoelement feature. They can also be found hybridized with Fabry–Perot radiation modes, trapped waveguide, or Bloch modes.

3.4.2 Some Numerical Modeling Results

We now examine some numerical modeling results for the nanoelement and planar structures of Figure 3.5. These results have been obtained using nanocrystalline silicon (nc-Si) as the absorber. For a fair comparison of the absorption capabilities, the same nc-Si effective thickness t' = absorber volume/unit cell area has been used in all cases where t' = 403 nm. The results to be presented are in the form of nc-Si absorption and electric field distributions as a function of (air) wavelength. Figure 3.7 gives the absorptance plots $A(\lambda)$ for the planar architecture with two different ARC layer thicknesses and also for the nanoelement architecture. The results are obtained by numerically solving Maxwell's equations for these geometries and materials [14]. If desired, the resulting absorption response $A(\lambda)$ can be converted to a short circuit current density J_{sc} by using the external quantum efficiency (EQE) in the expression $J_{sc} = e \int \Phi(\lambda) A(\lambda)(EQE) d\lambda$ where $\Phi(\lambda)$ is the impinging and spectrum, the integration is over the incoming spectrum [1]. Figure 3.7 shows that the nanoelement architecture, which can have radiation modes, trapped modes, and hybridization of these involved, does a very good job in light trapping when compared to either planar structure. Of

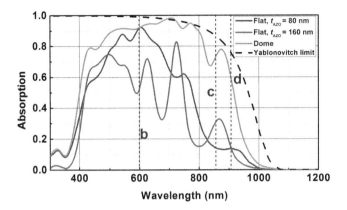

Fig. 3.7. The nc-Si A(λ) behavior for the planar architecture of Figure 3.5a for two TCO (AZO) coating thicknesses and for the nanoelement architecture of Figure 3.5b. The nc-Si Yablonovitch limit is shown for comparison. The various electric field distributions at the wavelengths denoted by (b), (c), and (d) are given in Figure 3.8.

course, the latter only have F-P radiation mode phenomena available for light trapping since there is no texturing or nanoelements to bring in other modes. The nanoelement architecture is seen to do a very good job even when compared to the Yablonovitch limit, which has ideal involvement of radiation and trapped traveling modes. As we have indicated, the nanoelement architecture, in principle, has a variety of modes to bring into play; i.e., all told, it offers the possibility of radiation modes, trapped traveling modes, trapped localized modes, and hybridization involvement in its light trapping. It also has the advantage of more involvement of radiation modes due to diffraction. Looked at this way, it is not surprising that a nanoelement architecture can do so well.

Figure 3.8a–c again shows the structures under discussion for convenience and Figure 3.8d–l give the electric field distributions present inside and around these three solar cells for different wavelength positions marked on the $A(\lambda)$ curves in Figure 3.7. To be specific, Figure 3.8d–f shows the three electric field distributions at the point marked (b) on the $A(\lambda)$ plot, Figure 3.8g–i shows the three electric field distributions at the point marked (c) on the $A(\lambda)$ plot, and Figure 3.8j–l shows the three electric field distributions at the point marked (d) on the $A(\lambda)$ plots. Importantly, as we noted in Chapter 1 (see Eq. 1.2), the photon density at any point in these distributions is proportional to the electric field value squared at the point. Of course, this photon density for a given λ determines $A(\lambda)$

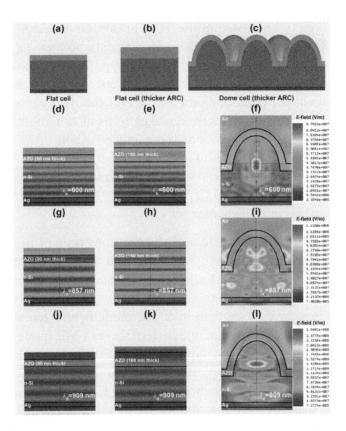

Fig. 3.8. The electric field distributions in the three devices of Figure 3.7 for λ = 600, 850, and 910 nm. Very different distributions are seen for the nanoelement architecture. The electric field strength scale is given on the figure.

for that wavelength provided the photon energy, which is more than or equal to the absorption threshold.[4] These electric field plots immediately allow us to see where the photons are and which mode or modes are involved in a particular light-trapping situation. Most importantly, they allow us to see why $A(\lambda)$ is superior for the nanoelement architecture.

For example, if we look first at the electric field distribution present in the two planar cells for $\lambda = 600$ nm, we can see the characteristic interference patterns of radiation F-P interactions in both. Interestingly, these patterns are seen to result in a stronger $A(600$ nm$)$ in Figure 3.7 for the 60 nm ARC since much of the electric field (i.e., much of the photon

---

[4] This threshold energy is the energy band gap for electrons and the excitation energy for excitons.

density) is seen to be wasted in the TCO in the 160 nm ARC case. The thinner TCO stands out at this 600 nm wavelength because, as discussed earlier, this ARC thickness is close to satisfying the ¼ wavelength ARC condition. Looking now at the electric field distribution present in the two planar cells for conditions (c) and (d) on the $A(\lambda)$ curves, we can again see the characteristic interference patterns of radiation F-P interactions, as we must since this mode is the only possibility. At $\lambda = 850$ nm these interference patterns result in a stronger $A(\lambda)$ in Figure 3.7 for the 160 nm TCO cell. At the longest wavelength explored $\lambda = 910$ nm radiation mode interference is seen to produce a very similar pattern for both TCO ARC values in these planar cells.

Looking now at the electric field distributions present in the nano-element architecture of Figure 3.7b, it is seen that these distributions are very different from those found in the planar architectures. At each wavelength explored some type of hybridization is present. In the case of $\lambda = 600$ nm, we can see there is a Mie mode hybridized with a F-P radiation mode. At $\lambda = 850$ nm, the pattern shows a Mie mode again but this time hybridized with what appears to be a trapped traveling mode. At the longest wavelength explored ($\lambda = 910$ nm), the electric field distribution shows a Mie mode hybridized with a F-P radiation mode. The electric field scale for these distributions shows that the highest fields lie in the nc-Si dome nanoelements for all three wavelengths. The distribution present for $\lambda = 600$ nm is seen to give the best $A(\lambda)$ among those of the other exemplarity wavelengths. This brief summary of some results for the nanoelement structure of Figure 3.7b gives us insight into the design flexibility and multiplicity of light-trapping approaches possible with this architecture.

For a guided mode or a Bloch mode to be populated, incoming light diffraction off an array has to have changed the photon momentum according to $k_{\parallel} = k_0 \sin\varphi + (2\pi m/a)$. Since impingement was modeled to have occurred with $\varphi = 0$ for Figure 3.8, Figure 2.10b shows Bloch modes cannot be launched in this case; i.e., all the possible $k_{\parallel} = (2\pi m/a)(m = \pm 1, \pm 2, \ldots)$ that can be imparted by diffraction lie in a part of the dispersion relationship that has no Bloch modes. Interestingly, if the incoming light impinges at an angle, then this same discussion, appropriately modified, also argues that Bloch modes cannot be launched, if they are present. When a nanoelement architecture has proportions that make it more appropriately thought of as a photonic crystal instead of as an array sitting on a waveguide, Bloch modes must be available. While we just argued that light incoming at any angle cannot couple to these Bloch modes, coupling may occur in a given structure for two possible reasons: (1) the momentum requirement analysis done above assumes array features which are much smaller than the wavelength of the incoming light and (2) Bloch wave involvement may be accomplished by hybridization rather than by diffraction. One could argue that Figure 3.8i gives an electric field distribution for the nanoelement structure that could be interpreted as a Bloch wave, rather than a guided wave, hybridized to the Mie mode.

Nanoelement architecture obviously can be extremely interesting. It can also be extremely varied. Figure 3.9a and b shows just two additional possibilities. In these particular examples, a nanofeature is found in each dome. In Figure 3.9a, this feature is a column whereas in Figure 3.9b it is a cone. As indicated in the figure, the materials used can also be varied.

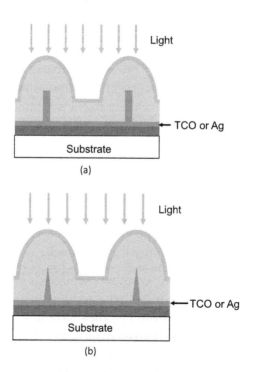

Fig. 3.9. Some other nanoelement architectures using domes. (a) Domes containing column nanofeatures and (b) domes containing cone nanofeatures.

These feature and material differences affect light trapping, the $A(\lambda)$ behavior, and can affect charge collection, as well [46].

3.5 STRUCTURES WITH PLASMONIC EFFECTS

In our initial discussion of spatially localized modes and light trapping, we listed Mie modes and localized surface plasma resonance (LSPR) modes as spatially localized states. We have seen that Mie states can contribute to light trapping and absorption directly if located in or near an absorber material and can contribute through hybridization with a radiation mode or a trapped traveling mode. Plasmonic modes are similar. We saw with Figure 3.3 that LSPR modes can contribute through scattering in a manner analogous to scattering from surface roughness. In addition, LSPR modes at metal particles can cause an intense near field in their vicinity which can increase absorption due to increased photon density and due to increased absorption arising from electric field intensity dependence [18]. Further, LSPR at metal nanoelements in a periodic array can also (1) diffract light into guided modes in an absorber and (2) hybridize with guided modes to produce surface plasmon polariton (SPP) waves [47,48]. Up to now, all the light-trapping schemes we have explored involved dielectric and absorber materials – perhaps with metal reflectors present. In this section, we take a closer look at what plasmonic effects can contribute and at metal nanoelement/metal nanoparticle light-trapping enhancement schemes.

3.5.1 Discussion

The LSPR behavior exhibited by metal nanoparticles and nanostructures can be excited by incident light of arbitrary impingement angle [44]. For this reason, the LSPR response in metal nanoparticles and nanostructures can induce a rich response and has raised a great deal of interest in its application to solar cells and detectors [44,45]. As we have noted earlier, loss issues can hinder this approach.

Surface plasmon polaritions, coupled oscillations of the free electron gas density in the metal with a guided electromagnetic wave in the absorber, have a dispersion relation; i.e., they have $\omega = \omega(k_{\parallel})$ rules. This dispersion relationship can be derived from Maxwell's equations [47,48]. In contrast to LSPR resonances, this dispersion relation shows that incoming light has difficulty coupling to these SPP modes unless there is a photon momentum adjustment. We have come to appreciate that an

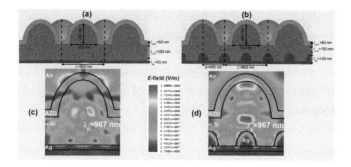

Fig. 3.10. A domed nc-Si nanoelement periodic array (a) without and (b) with a metallic nanoelement array at the back. Also shown in (c) is the electric field distribution at 967 nm for structure (a) and in (d) is the electric field distribution at 967 nm for structure (b).

effective means of addressing this problem is the presence of some sort of a repetitive array, which allows diffraction.

3.5.2 Some Numerical Modeling Results

Figure 3.10 gives numerical modeling results for electric field distributions for two nc-Si nanoelement architecture solar cells, one without (Figure 3.10a) and one with (Figure 3.10b) the addition of a silver back nanostructure array. The field plots have been obtained by solving Maxwell's equations for the cells. An example of plasmonic resonances and their hybridization possibilities may be seen in Figure 3.10d. This pattern of Figure 3.10d suggests SPP mode hybridization with a Mie mode, at least for $\lambda = 967$ nm. Figure 3.11 shows this hybridization is

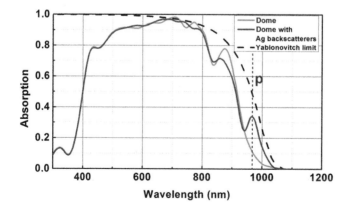

Fig. 3.11. The absorption as a function of wavelength for architectures (a) and (b) of Figure 3.10 along with the nc-Si Yablonovitch limit. The absorption at 967 nm is denoted by "p."

effective at enhancing the absorption at this wavelength. Back plasmonic array effects are showing up in Figure 3.11 at long wavelengths since such wavelengths can significantly permeate into the cell's back region. Studies have shown that plasmonic responses can also be effective at scattering a significant amount of incident light into guided modes in properly designed architectures [49].

CHAPTER 4

Summary

4.1 THE CURRENT PICTURE

In the case of planar structures, light trapping is principally limited to utilization of the radiation modes in the absorber material. These are always present. Overcoming this limitation of one trapping mechanism in a planar structure can be done by employing the roughening of surfaces, interfaces, or both to scatter energy into the trapped traveling modes also present in the absorber of a planar structure.

Nanoelement architectures, defined here as having features of the order of or smaller than the light wavelengths involved, solve this problem of limited light storage mechanisms by broadening the variety of optical phenomena and photon states present in a structure. This is accomplished by using nanoelement architectures to add nanoelement arrays and features to the structural layout. These arrays and features can be composed of dielectric (here including semiconductor and TCO materials) or metallic materials. The addition of periodic arrays, parallel to the entry plane, somewhere in the structure opens the door to diffraction. This optical phenomenon changes photon momentum parallel to the entry plane and can thereby systematically scatter photons into additional radiation modes in the absorber. These arrays also can scatter photons into the trapped traveling modes, which may otherwise be unavailable. The actual diffraction process can be straightforward diffraction at a dielectric periodic array or diffraction via plasmonic reradiation at a nanoparticle or nanostructured metallic arrays. Involving nanoelement architectures also directly introduces new, additional modes, which we term trapped localized modes. In the case of dielectric features and arrays these trapped localized states are Mie modes. These are characterized by their electric field distribution and symmetry, which mimic the features supporting them. In the case of metallic features and arrays these trapped localized states are localized surface plasmonic modes. All told, nanoelement architectures can have radiation modes,

Introduction to Light Trapping in Solar Cell and Photo-detector Devices. DOI: 10.1016/B978-0-12-416649-3.00004-X

trapped traveling modes (waveguide and Bloch modes), and trapped localized modes (Mie and plasmonic modes) as well as hybridizations of these. Depending on the relative sizes of the absorber thickness and feature dimensions, nanoelement architectures can be thought of as being an array of nanoelements positioned on a waveguide, as a photonic crystal, or as some intermediate.

4.2 SOME FUTURE DIRECTIONS?

There is a great deal of current interest in looking into whether random features or nanoelement periodic arrays can ultimately give better light-trapping performance [50,51]. Some of these results show that the nanoelement periodic array approach is better. Some show that disordered perturbations of the periodic array approach yield somewhat superior performance. From an engineering perspective, one might argue that making carefully controlled periodic arrays is a more manufacturable approach than making carefully controlled random structures – especially if a mistake here may actually be advantageous. One can also argue that a periodic array structure is better for carrier collection, the other key component of performance.

4.3 OVERVIEW

Planar solar cell and detector structures have radiation and trapped traveling modes. Unless texturing is used, the trapped traveling modes are largely inaccessible for use in trapping light. Nanoelement solar cell and detector structures also have radiation and trapped traveling modes. In addition they possibly have Bloch modes and they have trapped localized modes arising from dielectric or metallic features as well as the possibility of hybridization among modes. Nanoelement solar cell and detector architectures can have periodic arrays, which support diffraction. This feature allows the selected scattering ($k_{\parallel}$ change) needed to permit the use of trapped traveling modes. It also gives rise to additional accessible radiation modes. Unlike random texturing, however, the accessibility of trapped traveling modes depends on the angle of impingement for a given frequency. In the case of dielectric array features, this diffraction is simple cooperative scattering. In the case of metallic array features, diffraction involves plasmonic reradiative processes. Nanoelement architectures have a further unique aspect: features that can support localized modes. In the dielectric feature case, the additional states are Mie modes, which can have intense electric field distributions in the absorber, hybridizations with other modes, or both. In the metallic feature case, these are plasmonic modes, which can also have intense electric field distributions in the absorber, hybridizations with other modes, or both.

APPENDIX A

Yablonovitch Limit Derivation

A.1 INTRODUCTORY COMMENTS

We noted in Section 1.3.2.1 that ideal diffuse reflection at textured surfaces of an absorber material can trap light and thereby enhance absorption. An analysis by Yablonovitch using geometrical (ray) optics showed this enhancement can be as much as $4n^2$ where n is the absorber's index of refraction [11]. Since geometrical optics is used, it is inherently assumed that the thickness of the absorber t is greater than the wavelength λ of the radiation, which may not be true in some light-trapping structures. In this appendix, we derive this $4n^2$ factor following the approach of Ref. [11]. In so doing, we will take the opportunity to explore the interpretations, implications, and further assumptions behind this $4n^2$ factor. Unlike the approach of Ref. [11], we will not include transmission or reflection losses at interfaces in our derivation in order to follow a more direct path to this $4n^2$ result.

A.2 DERIVATION

Figure A.1 shows light of an intensity I_{ei} (energy/time–area) impinging normally from air onto an area element dA_s of an absorber's surface. Here the subscripts on I_{ei} stand for external impingement. The quantity I_{ei} may be a function of the wave angular frequency ω. If the actual incoming light intensity is I_{in} and it is impinging onto dA_s at some angle γ with respect to the normal of dA_s as seen in Figure A.1, then the normal intensity I_{ei} that we are going to use in this derivation is related to I_{in} by the ratio of the projected differential area "seen" by I_{in} to dA_s; i.e., $I_{ei} = \cos\gamma \, I_{in}$. As will unfold, we will do just fine by utilizing I_{ei} and we will not need to involve I_{in}.

This I_{ei} sets up an intensity B per steradian (energy/time–area–solid angle) inside the volume of the light-trapping structure for every solid angle Ω at dA_s. The quantity B adjusts until the light intensity escaping back out to air and the total absorption A inside the absorber (with

Introduction to Light Trapping in Solar Cell and Photo-detector Devices. DOI: 10.1016/B978-0-12-416649-3.00005-1

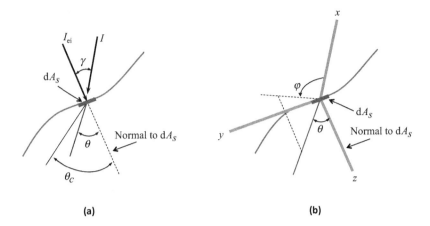

Fig. A.1. (a) Some area element dA_s is seen on the surface of a light-trapping region. The effective intensity I_{ei} is impinging normally on this element. (b) The spherical coordinate system used in the evaluation of the various integrations. A Cartesian coordinate system is shown for orientation.

units of energy/time–area) each come to steady state. This steady-state value of B is basic to our derivation. It and its units of intensity per solid angle may seem to be a strange invention but this definition of B from Ref. [11] allows us to (1) account for the total randomization of the light ray directions inside our structure caused by the multiple diffuse reflections and to (2) account for total internal reflection at dA_s which will limit the rays that can escape back out to air. The point of (2) is that B will allow us to account for the fact that not every ray headed toward dA_s can escape. Only those in an escape cone can do so successfully.

With this understanding of B, the product $B\, d\Omega$ is seen to be the intensity from the solid angle element $d\Omega$ heading toward our elemental area dA_s in Figure A.1a. Because of the multiple diffuse reflections and their randomizing affect, we assume B is not a function of the solid angle Ω. Since this is a derivation based on the ray picture of geometrical optics, there is no attempt to account for any impact on B of the physical optics phenomena of Chapter 1.

The escape cone mentioned above is not fully drawn in Figure A.1 for simplicity but only indicated by θ_c, a polar angle of the spherical coordinate system seen in Figure A.1b. The quantity θ_c defines the cone limiting the escape of B since

$$n \sin \theta_c = 1 \qquad (A.1)$$

Refraction sets this limit and any B impacting dA_s at angles $\theta \geq \theta_c$ cannot escape back into air from the absorber due to total internal reflection. With this picture, we can use the symbol I_{ie} to represent the intensity escaping through dA_s to air from inside the absorber and note that it must be given by

$$I_{ie} = \int_{\substack{\text{inside} \\ \text{the cone}}} B \cos\theta \sin\theta d\theta \, d\varphi \tag{A.2}$$

The relation $d\Omega = \sin\theta \, d\theta \, d\varphi$ has been used in Eq. A.2 to express the solid angle differential element $d\Omega$ in terms of the spherical coordinate system seen in Figure A.1b. The $\cos\theta$ in Eq. A.2 accounts for the projection that dA_s presents to an incoming B. As may be deduced from Figure A.1a and b, to cover the complete solid angle inside the escape cone the range of integration necessary for the azimuthal angle φ is $0 \leq \varphi \leq 2\pi$ whereas the range of integration necessary for the polar angle θ is $0 \leq \theta \leq \theta_c$. Using these limits, Eq. A.2 integrates to

$$I_{ie} = \frac{\pi B}{n^2} \tag{A.3}$$

While Eq. A.3 will be useful, it is not our goal. Instead our goal is to find out how much the trapping of light due to the incessant diffuse reflections going on inside the absorber effects absorption; i.e., our end goal is to determine the absorption A (expressed as energy/time–area) taking place in the absorber and ultimately find out where that $4n^2$ factor of Section 1.3.2.1 comes from.

To now determine A, we first note that, in general [1]:

$$dAbs = \alpha \, I \, dV \tag{A.4}$$

where dAbs is the absorption (energy/time–vol) in some volume element dV. In this general statement, I is some light intensity and α is the absorption coefficient for this light of angular frequency ω. In specifically formulating A, we first start with A', the absorption (energy/time–area) suffered by the rays B impinging on dA_s, and associate this with the volume $t\,dA_s$ where t is the thickness of the absorber. Hence the absorption produced by all the rays coming toward dA_s is

$$A' = \alpha t \int_0^{2\pi} \int_0^{\pi/2} B\sin\theta \; d\theta \; d\varphi \qquad (A.5)$$

The limits in Eq. A.5 come from this requirement of accounting for all rays heading toward dA_s. Equation A.5 integrates to

$$A' = 2\pi \; \alpha t B \qquad (A.6)$$

which gives

$$A = 4\pi \; \alpha t B \qquad (A.7)$$

where the factor of two difference between A and A' results from including rays moving toward and away from dA_s. In formulating Eq. A.5, we have effectively (1) once again neglected any physical optics effects and (2) assumed $t < \alpha^{-1}$, which either means the absorber is thin or the absorption coefficient is small or both.

The principle of detailed balance can now be used to determine A in terms of I_{ei}. It says that, in steady state, we must have

$$I_{ei} = I_{ie} + A \qquad (A.8)$$

Utilizing Eqs. A.3 and A.7 in Eq. A.8 and solving for A gives

$$A = 4n^2 \left[\frac{\alpha t I_{ei}}{4n^2 \alpha t + 1} \right] \qquad (A.9)$$

The absorptance, defined by A/I_{ei}, may be written using Eq. A.9 as

$$\text{Absorptance} = 4n^2 \left[\frac{\alpha t}{4n^2 \alpha t + 1} \right] \qquad (A.10)$$

From its definition, we note that absorptance must be ≤ 1. Equation A.9 can be approximated by

$$A = 4n^2 \left[\alpha t I_{ei} \right] \qquad (A.11)$$

If $4n^2\,\alpha t \ll 1$. In that case

$$\text{Absorptance} = 4n^2\alpha t \qquad\qquad (A.12)$$

The quantity $\alpha t I_{ei}$ is the absorption per area that would take place in an absorber with $t < \alpha^{-1}$ if there were no reflection of any kind inside the structure. Consequently, Eq. A.11 says this absorption can be increased by the factor $4n^2$ if an ideal textured absorber were in use that had ideal diffuse reflection at both surfaces. All the assumptions made in deriving these equations including those specifically needed for Eq. A.11 must be kept in mind when employing the $4n^2$ Yablonovitch factor of Eq. A.11. As stated in Section 1.3.2.1, this $4n^2$ factor provides a straightforward, useful limit for absorption enhancement assessment. However, as should be clear from this discussion, it is a "limit" based on a number of assumptions.

Fresnel Equations for the Situation of Section 2.2

B.1 BACKGROUND ON THE FRESNEL EQUATIONS USED IN SECTION 2.2

In Section 2.2, we established three Fresnel equations. These three come from the standard Maxwell's equations boundary condition

$$\int_s \nabla \times \overline{\xi}(\overline{r},t)\,dA = -\int_s \frac{\partial \overline{B}(\overline{r},t)}{\partial t}\cdot dA = \oint_c \overline{\xi}(\overline{r},t)\cdot dl \qquad \text{(B.1)}$$

which, in the situation being modeled, implies tangential electric field components must be preserved at each boundary. Here ξ is the electric field and B the magnetic field density. With the polarization assumed, imposing this condition at the three boundaries leads to the three equations given in Section 2.2.

Similarly, the standard Maxwell's equations boundary condition

$$\int \nabla \times \overline{H}(\overline{r},t)\cdot dA = \oint_c H(\overline{r},t)\cdot dl \qquad \text{(B.2)}$$

implies in the situation being modeled in Section 2.2 that the magnetic tangential components must also be preserved. Here H is the magnetic field. Imposing this condition at the three boundaries leads, with the polarization assumed and with some further manipulation, to three more boundary condition equations giving a determined mathematical system with six equations and six unknowns.

Introduction to Light Trapping in Solar Cell and Photo-detector Devices. DOI: 10.1016/B978-0-12-416649-3.00006-3

Index of Refraction, Permittivity, and Absorption Coefficient

C.1 THE RELATIONSHIPS

The relationships among the index of refraction n, the permittivity $\varepsilon\varepsilon_0$, and the absorption coefficient α need to be clearly defined to facilitate various discussions in the text. Here the quantity ε is the dielectric constant and ε_0 is the permittivity of free space. A first point to make is that while α is a real number, n and ε are parts of the complex quantities [16]; i.e., the complex index of refraction is $\tilde{n} = n(\omega) + in_i(\omega)$ and the complex dielectric constant is $\tilde{\varepsilon} = [\varepsilon(\omega) + i\varepsilon_i(\omega)]$. The complex permittivity can be related to the complex index of refraction by first writing the definition of the index of refraction, i.e.,

$$n = \frac{\text{speed of light in vacuum}}{\text{speed of light in medium}} = \left[\frac{(\varepsilon\varepsilon_0\mu\mu_0)}{(\varepsilon_0\mu_0)}\right]^{\frac{1}{2}} \quad \text{(C.1)}$$

In Eq. C.1 $\mu\mu_0$ is the permeability, μ_0 the permeability of free space, and μ the relative permeability, which is essentially unity for many materials. Using that information and the complex definitions of index of refraction and permittivity then gives

$$[n(\omega) + in_i(\omega)]^2 = [\varepsilon(\omega) + i\varepsilon_i(\omega)] \quad \text{(C.2)}$$

This equation shows that the conversion between these various real and imaginary quantities is given by Ref. [16]

$$\varepsilon(\omega) = n^2 - n_i^2(\omega) \quad \text{(C.3)}$$

$$\varepsilon_i(\omega) = 2nn_i(\omega) \quad \text{(C.4)}$$

Introduction to Light Trapping in Solar Cell and Photo-detector Devices. DOI: 10.1016/B978-0-12-416649-3.00007-5

$$n = \sqrt{\frac{\sqrt{\varepsilon(\omega)^2 + \varepsilon_i(\omega)^2} + \varepsilon(\omega)}{2}} \qquad (C.5)$$

$$n_i = \sqrt{\frac{\sqrt{\varepsilon(\omega)^2 + \varepsilon_i(\omega)^2} - \varepsilon(\omega)}{2}} \qquad (C.6)$$

The relation between these quantities and the absorption coefficient α may be determined by writing the expression for a plane wave $e^{i(\tilde{n}k_0 x - \omega t)}$ moving in a medium with a complex index of refraction $\tilde{n}$ where k_0 is the wave number in air. This becomes

$$e^{i(\tilde{n}k_0 x - \omega t)} = e^{i((n(\omega) + in_i(\omega))k_0 x - \omega t)} = e^{i(n(\omega)k_0 x - \omega t)} e^{-k_0 n_i(\omega)x} \qquad (C.7)$$

Since the intensity of this light wave is given by Eq. C.7 times its complex conjugate, the intensity is seen from Eq. C.7 to decay as $e^{-2k_0 n_i(\omega)x}$. According to the definition of absorption coefficient, in an infinite medium light intensity decay should follow the Beer-Lambert law $e^{-\alpha x}$ [1]. This allows us to make the association that

$$\alpha = 2k_0 n_i(\omega) \qquad (C.8)$$

REFERENCES

[1] Fonash SJ. Solar cell device physics. 2nd ed. New York, NY: Academic Press, Inc; 2010.

[2] Homola J, Yee SS, Gauglitz G. Surface plasmon resonance sensors: review. Sens Actuat B54 1999;3–15.

[3] Abdulhalim I, Zourob M, Lakhtakia A. Surface plasmon resonance for biosensing: a mini-review. Electromagnetics 2008;28:214–42.

[4] Nishijima Y, Ueno K, Kotake Y, Murakoshi K, Inoue H, Misawa H. Near-infrared plasmon-assisted water oxidation. J Phys Chem Lett 2012;3(10):1248–52.

[5] Ingram DB, Linic S. Water splitting on composite plasmonic-metal/semiconductor photoelectrodes: evidence for selective plasmon-induced formation of charge carriers near the semiconductor surface. J Am Chem Soc 2011;133(14):5202–5.

[6] Nagpal P, Han SE, Stein A, Norris DJ. Efficient low-temperature thermophotovoltaic emitters from metallic photonic crystals. Nano Lett 2008;8(10):3238–43.

[7] Stuart HR, Hall DG. Thermodynamic limit to light trapping in thin planar structures. J Opt Soc Am A 1997;14:3001.

[8] Feynman F. The Feynman lectures on physics: commemorative issue. Reading, MA: Addison-Wesley; 1989.

[9] Hecht E. Optics. 4th ed. London, UK: Addison-Wesley Publishing; 2002.

[10] Smith WJ. Modern optical engineering. 4th ed. New York, NY: McGraw-Hill Companies; 2007.

[11] Yablonovitch E. Statistical ray optics. J Opt Soc Am 1982;72(7):899–907.

[12] Yu Z, Raman A, Fan S. Fundamental limit of light trapping in grating structures. Opt Exp 2010;18(S3):A366–80.

[13] Lipson SG, Lipson H, Tannhauser DS. Optical physics. 3rd ed. Cambridge, NY: Cambridge University Press; 1995.

[14] http://www.ansys.com/Products/Simulation+Technology/Electronics/Signal+Integrity/ANSYS+HFSS

[15] Kumar S, Sharma TP, Zulfequar M, Husain M. Characterization of vacuum evaporated PbS thin films. Phys B Conden Matter 2003;325:8–16.

[16] Kittel C. Introduction to solid state physics. 8th ed. Hoboken, NJ: John Wiley & Sons, Inc; 2005.

[17] Haug F-J, Soderstrom T, Cubero O, Terrazzoni-Daudrix V, Ballif C. Plasmonic absorption in textured silver back reflectors of thin film solar cells. J Appl Phys 2008;104:064509.

[18] Murai S, Verschuuren MA, Lozano G, Pirruccio G, Rodriguez SRK, Gomez Rivas J. Hybrid plasmonic-photonic modes in diffractive arrays of nanoparticles coupled to light-emitting optical waveguides. Opt Exp 2013;21(4):4250–62.

[19] Bhattacharya J, Chakravarty N, Pattnaik S, Slafer WD, Biswas R, Dalal VL. A photonic-plasmonic structure for enhancing light absorption in thin film solar Cells. Appl Phys Lett 2011;99:131114.

[20] Ferry VE, Verschuuren MA, Li HBT, Verhagen E, Walters RJ, Schropp REI, et al. Light trapping in ultrathin plasmonic solar cells. Opt Exp 2010;18(S2):A237–45.

Introduction to Light Trapping in Solar Cell and Photo-detector Devices. DOI: 10.1016/B978-0-12-416649-3.00008-7

[21] Biswas R, Zhou D, Curtin B, Chakravarty N, Dalal V. Surface plasmon enhancement of optical absorption of thin film a-Si:H solar cells with metallic nanoparticles. 34th IEEE PVSC Proceedings; Philadelphia, PA, June 7–12, 2009. p. 000557–60.

[22] Chou SY, Ding W. Ultrathin, high-efficiency, broad-band, omniacceptance, organic solar cells enhanced by plasmonic cavity with subwavelength hole array. Opt Exp 2013;21:A60.

[23] Orfanidis SJ. Electromagnetic wave and antennas. http://eceweb1.rutgers.edu/~orfanidi/ewa/; 2014. p. 153–185.

[24] Joannopoulos JD, Johnson SG, Winn JN, Meade RD. Photonic crystals: molding the flow of light. 2nd ed. Princeton, NJ: Princeton University Press; 2008.

[25] Li ZY, Ho KM. Application of structural symmetries in the plane-wave-based transfer-matrix method for three-dimensional photonic crystal waveguides. Phys Rev B 2003;68:245117.

[26] Haus HA. Waves and fields in optoelectronics. Englewood Cliffs, NJ: Prentice-Hall; 1984.

[27] Yang Q, Zhang XA, Bagal A, Guo W, Chang C. Antireflection effects at nanostructured material interfaces and the suppression of thin-film interference. Nanotechnology 2013;24:235202.

[28] http://www.pveducation.org/pvcdrom/design/anti-reflection-coatings

[29] Kishino K, Unlii MS, Chyi J, Reed J, Arsenault L, Morkoq H. Cavity-enhanced (RCE) photodetectors. IEEE J Quant Elect 1991;21:8. 2025.

[30] Ferry VE, Munday JN, Atwater HA. Design considerations for plasmonic photovoltaics. Adv Mater 2010;22:4794–808.

[31] Campbell P, Green MA. Light trapping properties of pyramidally textured surfaces. J Appl Phys 1987;62:243–9.

[32] Werner JH, Bergmann R, Brendel R. The challenge of crystalline thin film solar cells Festkörperprobleme 34. Adv Solid State Phys 1994;34:115–46.

[33] Hegedus SS, Deng X. Analysis of optical enhancement in a-Si n-i-p solar cells using a detachable back reflector. Record 25th IEEE Photovoltaics Specialists Conf.; 1996. p. 1061–4.

[34] Benagli S, Borrello D, Vallat-Sauvain E, Meier J, Kroll U, Hoetzel H, et al. High-efficiency amorphous silicon devices on LPCVD-ZnO TCO prepared in industrial KAI-M reactor. In: Hamburg, Germany. Proceedings 24th European Photovoltaic Solar Energy Conf.; 2009.

[35] Ruby DS, Zaidi SH, Narayanan S, Damiani BM, Rohatgi A. RIE-texturing of multicrystalline silicon solar cells. Sol Energ Mater Sol Cells 2002;74:133–7.

[36] Nam WJ, Liu T, Wagner S, Fonash SJ. A study of lateral collection single junction a-SI:H solar cell devices using nanoscale columnar array structures. 35th IEEE Photovoltaics Specialists Conf. (PVSC); 2010. p. 923.

[37] Zhu J, Hsu C, Yu Z, Fan S, Cui Y. Nanodome solar cells with efficient light management and self-cleaning. Nano Lett 2010;10:1979.

[38] Atwater HA, Polman A. Plasmonics for improved photovoltaic devices. Nat Mater 2010;9: 205–13.

[39] Nam WJ, Benanti TL, Varadan VV, Wagner S, Wang Q, Nemeth W, et al. Incorporation of a light and carrier collection management nano-element array into superstrate a-Si:H solar cells. Appl Phys Lett 2011;99:073113.

[40] Battaglia C, Escarre J, Söderström K, Charrière M, Despeisse M, Haug FJ, et al. Nanomoulding of transparent zinc oxide electrodes for efficient light trapping in solar cells. Nat Photonics 2011;5:535.

[41] Haug FJ, Söderström T, Cubero O, Terrazzoni-Daudrix V, Ballif C. Influence of the ZnO buffer on the guided mode structure in Si/ZnO/Ag multilayers. J Appl Phys 2009;106:044502.

[42] Brongersma M, Cui Y, Fan S. Light management for photovoltaics using high-index nano-structure. Nat Mater 2014;13:451.

[43] Mie G. Beiträge zur optik trüber medien, speziell kolloidaler metallösungen. Ann Phys 1908;330(3):377–445.

[44] Maier SA, Atwater HA. Plasmonics: localization and guiding of electromagnetic energy in metal/dielectric structures. J Appl Phys 2005;98:011101.

[45] Dahmen C, von Plessen G. Optical effects of metallic nanoparticles. Aust J Chem 2007; 60(7):447–56.

[46] Nam WJ, Ji L, Varadan VV, Fonash SJ. Exploration of nano-element array architectures for substrate solar cells using an a-Si:H absorber. J Appl Phys 2012;111:123103.

[47] Kreibig U, Vollmer M. Optical properties of metal clusters. Springer Series in Materials Science (Book 25). Berlin, Germany: Springer-Verlag; 1995.

[48] Raether H. Surface plasmons on smooth and rough surfaces and on gratings. Springer Tracts in Modern Physics (Volume 111). Berlin, Germany: Springer-Verlag; 1988.

[49] Paetzold UW. Light trapping with plasmonic back contacts in thin-film silicon solar cells. Doctoral dissertation. RWTH Aachen University. Retrieved from http://darwin.bth.rwth-aachen.de/opus3/volltexte/2013/4602/.

[50] Battaglia CM, Hsu CM, Söderström K, Escarré J, Haug FJ, Charrière M, et al. Light trapping in solar cells: can periodic beat random? ACS Nano 2012;6(3):2790–7.

[51] Pereeti R, Gomard G, Lalouat L, Seassal C, Drouard E. Absorption enhancement in photonic crystal thin films by pseudo disordered perturbations. Mater Res Soc Symp Proc 2014;1627.

Printed and bound by CPI Group (UK) Ltd, Croydon, CR0 4YY

03/10/2024

01040421-0006